RESILIENT URBAN DRAINAGE SYSTEM STRATEGIES FOR EXTREME WEATHER

DESIGN FOR THE "NEW NORMAL"

Dr. Thewodros Geberemariam, Ph.D., P.E., BCEE, D.WRE

Library of Congress Cataloging-in-Publication Data can be found in the WAV section of the publisher's website at www.jrosspub.com/wav.

Phone: (954) 727-9333
Fax: (561) 892-0700
Web: www.jrosspub.com

CONTENTS

PREFACE

Recent catastrophic floods—attributed to climate change—underscore the significant challenge that intense precipitation and extreme weather events pose to urban drainage infrastructure. Numerous hydrological studies have explored the potential impact of climate change on these systems, thereby predicting adverse consequences.

Current initiatives to ensure the resilience of urban drainage infrastructure heavily rely on standard design guidelines. These guidelines dictate how infrastructure systems are designed to withstand flooding. However, the increasing frequency and intensity of flooding due to significant storm events have raised questions about the future viability of these methodologies and standards. Thus, policymakers, planners, and design professionals are considering supplementing existing design standards, policies, and regulations to accommodate the "New Normal" of extreme weather.

Resilient Urban Drainage System Strategies for Extreme Weather aims to analyze design practices and strategies that are suitable for extreme weather conditions in the "New Normal" era by focusing on urban drainage infrastructures. It anticipates that *current* stormwater infrastructure and flood management drainage design criteria will remain essential elements of the design process of small watershed components/subsystems. However, at the system/larger watersheds level, it suggests that approaches such as fail-safe, safe-to-fail, and robust decision making (RDM) could enhance existing design approaches. These methods explicitly consider the consequences of failure in the design and risk analysis processes.

This book builds on the concept of resilient urban drainage system design solutions for "New Normal" extreme weather, chapter by chapter. It explores various approaches to improve drainage system design requirements, and each chapter concludes with objective questions that offer a variety of possible answers and real-world practical problems. Intended to be user-friendly, this book aims to foster an appreciation for supplementing existing drainage system design criteria through a simplified approach and underscores the increasing importance of adopting a multi-scalar perspective on resilience to address the escalating challenges faced by urban municipalities. It will be valuable to professionals in the field of drainage, graduate students pursuing their M.Sc. and Ph.D. degrees, and members of the academic community.

ABOUT THE AUTHOR

Dr. Thewodros K. Geberemariam, Ph.D., P.E., D.WRE, BCEE, EXW[SM], PMP, LSIT, QPSWPPP, QCIS, ENV SP, M. ASCE, AACPM, has over 25 years of extensive experience in urban infrastructure and water resources engineering. His expertise is designing, analyzing, modeling, and evaluating urban drainage systems and flood control structures. He has been responsible for designing major and minor stormwater drainage conveyance systems, as well as water, sewer, and street improvement projects. He holds a Ph.D. and M.Sc. in Civil Engineering with a specialization in Urban Infrastructure and Water Resources Engineering from the New York University Tandon School of Engineering and the National University School of Engineering and Computing, respectively.

He is a licensed professional engineer in California and New York. Additionally, he is a Board-Certified Environmental Engineer and has been recognized as a Diplomate, Water Resources Engineer by the American Academy of Water Resources Engineers, a Division of the American Society of Civil Engineers (ASCE). He has been certified by the American Council of Engineering Companies as an expert witness in surface water hydrology and hydraulics.

Over the years, he has taught undergraduate and graduate-level courses including hydrology, hydraulic engineering, fluid mechanics, engineering economics, and an introduction to highway design. He has published more than 50 advanced works on modeling, forecasting, asset management, hydrology, hydraulics analysis, and optimization.

Thewodros served as a peer reviewer and a technical advisor for ASCE and his toolbox spans from simple analytical models to the most detailed, complex, numerical, stochastic 3D models, as well as advanced machine learning models. He also has served as a lead technical expert on a wide range of water resources engineering projects, including hydrology and hydraulics analysis, surface hydrology, stormwater best management practices (BMP) design and implementation, water quality monitoring and evaluations, hydroclimate modeling, rainfall-runoff modeling, flood forecasting and control, watershed and floodplain management, and stormwater infrastructure risk analysis. Thewodros also consults on domestic and international water resources engineering projects.

Web
Added
Value™

This book has free material available for download from the
Web Added Value™ resource center at *www.jrosspub.com*

At J. Ross Publishing we are committed to providing today's professional with practical, hands-on tools that enhance the learning experience and give readers an opportunity to apply what they have learned. That is why we offer free ancillary materials available for download on this book and all participating Web Added Value™ publications. These online resources may include interactive versions of material that appears in the book or supplemental templates, worksheets, models, plans, case studies, proposals, spreadsheets and assessment tools, among other things. Whenever you see the WAV™ symbol in any of our publications, it means bonus materials accompany the book and are available from the Web Added Value Download Resource Center at www.jrosspub.com.

Downloads for *Resilient Urban Drainage System Strategies for Extreme Weather* include instructional material for classroom use (lecture slides, exercise solutions, etc.).

1

PURPOSE AND SCOPE

1.1 INTRODUCTION

Reliability, risk, and resilience are frequently discussed in drainage system design publications, especially in relation to climate change and increasing urbanization. However, a significant and under-researched disparity exists between theoretical knowledge and the practical implementation of drainage infrastructure by field professionals. Thus, many drainage specialists grapple with understanding a drainage system's response to the "New Normal" of today's extreme weather, which was not considered during its initial design phase.

This gap is further widened by the need to evaluate critical infrastructure for potential disasters of greater magnitude than initially anticipated. This book scrutinizes the relationship between the normalized capacity or drain down/emptying time of various drainage system components and the shape, intensity, duration, and spatial extent of storm events. Other significant factors discussed include climate change and densification. The discussion emphasizes the potential benefits of integrating natural systems and low-impact development or green infrastructure practices into current drainage system design criteria. The goal is to augment the system's capacity to manage extreme weather events, such as fail-safe, safe-to-fail, and robust decision making (RDM) in specific situations.

A crucial aspect of resilience measures in urban drainage infrastructure systems is the incorporation of design storm criteria. These criteria specify the intensity or frequency that the systems are built to withstand. Factors such as climate change and the increasing complexity of urban systems are challenging the sustainability of current methods and the implementation of design storm criteria.

This book seeks to identify design methodologies and approaches that are suitable for the operational environments of modern cities and infrastructure, which are becoming increasingly complex and dynamic. To effectively address the challenges faced by drainage infrastructure systems in large municipalities, it is essential to adopt a multi-scalar perspective on resilience. This approach will significantly contribute to addressing these issues comprehensively. It is expected that the inclusion of return periods (or similar criteria) will remain necessary during the design process for individual components or subsystems. The current methodologies can be enhanced by incorporating the consideration of failure effects into the design and management processes. Approaches such as safe-to-fail and RDM appear particularly apt for addressing the needs of the entire system(s).

1.2 DESIGN FOR THE "NEW NORMAL"

Coastal and rural villages are increasingly experiencing submersion in water. These extreme weather events and rising sea levels can be attributed to the impacts of climate change. Extreme precipitation and weather events pose a threat to urban infrastructure systems, as demonstrated by recent catastrophic flooding disasters. According to design professionals and policymakers, urban infrastructure design has had to adapt to the "New Normal" of extreme weather. A crucial aspect of resilience initiatives in urban and infrastructure systems is the accurate determination of the frequency and intensity of extreme weather events that these systems are designed to withstand.

This textbook aims to delineate effective design methods and strategies to address the challenges posed by extreme weather conditions on local municipalities, infrastructure, and cities. To effectively mitigate the escalating challenges faced by cities and infrastructure systems, it is crucial to adopt a holistic approach that integrates various resilience levels. The existing stormwater infrastructure and flood control drainage standards are expected to continue being vital considerations for design input at the individual component and subsystem levels. At the system level, techniques such as safe-to-fail and RDM seem highly appropriate for enhancing current practices by explicitly considering the impact of failures during the design and risk analysis stages.

Presently, urban drainage system design textbooks and stormwater runoff management guidelines lack guidance on accommodating increased runoff resulting from more intense storm events. This book aims to provide advanced instruction on the fundamental principles of resilience in urban stormwater management within the "New Normal" framework, thereby enhancing existing design standards to accommodate the challenges posed by increasingly severe weather conditions.

1.3 OBJECTIVES

This book has four primary objectives:

1. Delineate effective design methods and strategies to address the challenges posed by extreme weather conditions that are being experienced by local municipalities and city infrastructure in the "New Normal" era.

2. Assist design professionals in developing innovative solutions that align with the existing drainage system design standards for extreme weather conditions.

3. Enhance existing large-scale drainage systems and establish preliminary strategies such as safe-to-fail and RDM that consider the consequences of failure during the design and management phases.

4. Provide suggestions on design principles and methods for integrating extreme weather into urban drainage practices. The significant gap between theory and practical application in incorporating extreme weather into urban drainage practice needs to be highlighted. This gap motivates the need to evaluate elements like *essential infrastructure* for disasters that are

significantly more severe than previously imagined. This book aims to bridge this crucial gap using fundamental and straightforward methods.

1.4 DISTINCTIVE FEATURES OF THE BOOK

The distinctive features of this book include:

1. Adopting contemporary subjects, including RDM, safe-to-fail, and enhanced modeling and sensing techniques
2. Attempting to clearly incorporate failure effects into the design and management processes while demonstrating alternative solutions to supplement current design methodologies
3. Using simple and fundamental methods to bridge the critical design gap related to extreme weather
4. Employing a data-driven approach to differentiate between meteorological and climate factors that influence extreme rainfall
5. Primarily focusing on the effort to construct urban drainage systems that are resilient to extreme weather
6. Offering recommendations on risk assessment techniques to analyze the likelihood and effects of drainage excess
7. Detailing suggestions for layout and planning to mitigate the effects of drainage system overflow
8. Providing best practice advice for designing urban drainage systems capable of sustainably handling periods of excess

2

EXTREME WEATHER DUE TO
CLIMATE CHANGE

2.1 INTRODUCTION

Climate change poses numerous challenges to city municipalities. Consequently, governments should articulate a new vision of resilience and sustainability that aligns with the interests of economies, ecosystems, and communities in the era of "New Normal" extreme weather due to climate change.

The shifting climate influences the variability of weather and climatic events, resulting in unprecedented extremes in frequency, intensity, spatial extent, duration, and timing. Weather or climate occurrences can give rise to extreme circumstances or effects, even if they do not exhibit statistical exceptionalism. This can occur either by surpassing a critical threshold in a social, ecological, or physical system or by coinciding with other events. Certain climate extremes, such as droughts and floods, could potentially arise from the convergence of various weather or climate phenomena that may not exhibit significant extremity when considered individually. The potential impact of a weather system, such as a tropical cyclone, can vary depending on its landfall location and timing. Even if the storm's intensity is not exceptionally severe relative to other tropical storms, it can still have significant consequences. However, not all extremes necessarily result in negative consequences. The correlation between changes in extremes and changes in mean climate can be attributed to the prediction that future conditions for certain variables will fall within the margins of present-day conditions. Natural climate variability, encompassing events such as significant flooding, is responsible for numerous instances of extreme weather and climate conditions. Additionally, natural climate oscillations occurring over multiple years or decades provide a contextual backdrop for anthropogenic climate change. Even in the absence of anthropogenic climate change, a significant spectrum of natural weather and climatic extremes would persist.

Climate change and its far-reaching consequences represent some of the most pressing global challenges we will face this century. In recent years, there has been a rapid increase in the pace of change. Between 1995 and 2005, there was a notable 20% spike in the amount of atmospheric carbon dioxide. According to the Intergovernmental Panel on Climate Change (IPCC 2007), the same period has witnessed 11 of the highest recorded temperatures since data collection began in 1850. The hydrologic cycle, also known as the water cycle, is influenced by the increase in global mean temperatures. According to the United Nations IPCC, hydrologic change is classified as *highly probable*, leading to an increase in instances of heavy precipitation (2007). This transition will significantly impact infrastructures and urban landscapes. Urban areas worldwide have encountered numerous instances of flooding in recent years, primarily due to substantial precipitation. It is expected that there will be an increased occurrence of torrential downpours in the future, which may result in additional harm to individuals and infrastructure. Conducting an analysis of the potential impacts of temperature and precipitation changes on urban drainage systems, as well as exploring strategies for implementing flood mitigation measures, is of the utmost importance in preventing adverse consequences. The purpose of this chapter is to provide an overview of various techniques and research findings that are related to the simulation of altered precipitation effects on urban drainage systems.

2.2 MODELING CLIMATE CHANGE

Climate models are intricate data models that simulate planetary behavior by incorporating mathematical representations of the climate system and the interactions of its various components. These models integrate inputs from multiple emission scenarios, which encompass a range of assumptions, including factors such as population, energy demand, and land use. The measurement of radiatively significant gas emissions plays a pivotal role in these climate models.

Regional climate models (RCMs) leverage data from global models to provide a detailed analysis of specific geographical areas. The IPCC has proposed approximately 40 scenarios, which can be categorized into different scenario families. The Rossby Centre, a part of the Swedish Meteorological and Hydrological Institute, developed RCA3, the latest regional atmospheric climate model in Sweden. RCA3 incorporates inputs from the European Centre Hamburg Model Version 4 (ECHAM4). The RCA3 model generates output data for a simulation period of 140 years, with a spatial resolution of 50 by 50 kilometers and a temporal resolution of 30 minutes. Given its superior temporal resolution relative to most previous climate models, it is better equipped to capture short-term rainfall patterns.

Climate models are often used to represent emissions scenarios individually. A plausible scenario is one that holds credibility, even if it may not necessarily exhibit a significant likelihood. Each model run is based on a specific greenhouse gas (GHG) emissions scenario. Even when using the same GHG emission assumptions, different climate models will yield varying forecasts of changes in the global and regional climate. However, each run of a climate model can be considered a reasonable projection of potential climate change. Given the wide array of emissions scenarios and models available, experts can provide a set of climate change scenarios (climate model runs) as the most reliable means to predict potential impacts. Using a variety of scenarios to represent a reasonable range of climate-related uncertainty is a useful general guideline. It is important to record a broad range of crucial variables, such as temperature and precipitation.

Hence, it is advisable to consider scenarios that encompass a diverse range of potential changes in precipitation patterns, especially when addressing climate change concerns. However, obtaining valuable precipitation data from climate models can be challenging. Thus, assessments of climate change impacts on climatic variables, such as increased precipitation, often rely on simulations conducted using climate models, specifically atmosphere-ocean circulation models, including general circulation models (GCMs) and RCMs.

During specific time intervals of the global simulation, RCMs can utilize initial and boundary conditions derived from the output of GCMs, a process known as *dynamic downscaling*. Currently, there is no mechanism for transmitting input from the RCM simulation to the driving GCM, resulting in a unidirectional nature of this technique. The primary role of the GCM in this simulation methodology is to accurately represent the global circulation system's response to significant external stimuli. The RCM enhances the accuracy of simulating climatic variables at smaller spatial scales by incorporating finer-scale influences, such as topographical features. However, there is a lack of comprehensive understanding regarding the complex mechanisms responsible for precipitation formation, especially considering its generation at both fine spatial and temporal scales.

Incorporating these processes into regional and global climate models presents challenges for experts. The limitations of numerical stability and computational efficiency restrict the ability to address local, short-lived precipitation-generating mechanisms, consequently limiting the temporal and spatial scales that can be incorporated into models. Therefore, there is a current constraint on the extent to which dynamic downscaling can be employed while maintaining accurate outcomes. It is important to note that the anticipated intensities of severe precipitation often exhibit a systematic bias, specifically being underestimated, due to the process of dynamic downscaling.

Most RCM simulations are currently available with daily temporal resolution and spatial resolution ranging between 25 and 50 km. Additionally, certain RCM simulations are also available with hourly temporal resolution and a spatial resolution of 10 km. Occasionally, higher resolutions can be achieved by incorporating statistical downscaling techniques into dynamic simulations, allowing for the inclusion of a bias correction as an integral part of the RCM simulation. Some RCMs possess high resolution but may lack the capability to accurately represent the intricate surface dynamics within heterogeneous regions. In these circumstances, it is recommended to use data from lower resolution climate models along with an additional statistical downscaling step for a more effective approach.

2.2.1 The Process of Formulating Climate Change Scenarios

Climate change effects can be divided into two broad approaches, depending on the method used to determine the projected direction and potential magnitude of climate change in the specific area under study. One approach involves the use of artificial climate change scenarios, where the historical average temperature and precipitation are intentionally altered by predetermined amounts on an annual, seasonal, or monthly basis. This approach mitigates the inherent uncertainties associated with GCMs and enables the application of sensitivity analysis. Sensitivity analysis is a useful tool for assessing the magnitude of climate change required to trigger significant impacts. The model calculates the potential impact on a hydrological variable due to a sequence of incremental changes in a climatic variable.

The observed changes in climatic variables may not necessarily indicate the direct consequences of increased concentrations of GHGs in the atmosphere. This limitation represents an inherent disadvantage associated with the generation of synthetic scenarios. This issue can be addressed by determining the magnitudes of change based on relevant data rather than selecting them arbitrarily. This may involve considering variations in historical data or evaluating the range of changes predicted by RCMs. The level of variability in the scenario remains unchanged when the adjustments are applied to historical climate data. This raises a concern since the impact of climate change is anticipated to affect variability, particularly in relation to precipitation patterns. The daily scaling method was devised as a solution to address this concern by integrating change factors into the analysis of historical precipitation data. In this methodology, the factors of change are determined based on the proportional magnitude of the event rather than maintaining a constant value across all years, seasons, or months.

An alternative approach to developing climate change scenarios involves utilizing one or more GHG emission scenarios, typically sourced from the IPCC Special Report on Emissions Scenarios. GCMs employ these scenarios to drive extensive simulations of the interrelated ocean-atmosphere system, allowing for the prediction of the climate's reaction to the expected increase in GHG concentrations. To enhance the applicability of these models for hydrological applications, downsizing is required for their outputs. For this task, we can employ an RCM that takes into account local topography and other climate parameters or a statistical downscaling strategy that modifies past climate records to reflect anticipated future changes.

2.2.2 Analysis of Flooding and Disaggregation of Rainfall

Urban drainage models are crucial tools for assessing the impact of climate change on urban drainage systems. These models aid in understanding and predicting how climate change scenarios will affect the functioning of urban drainage systems. By using these models, we can generate estimates and projections that provide valuable insights into the potential consequences of climate change on urban drainage. The output time series of the climate model can be directly inputted into the drainage model to achieve this. Statistical downscaling is necessary for the urban drainage model.

Implementing regular corrections can help mitigate consistently divergent drainage outcomes in control simulations relative to those achieved after calibrating the drainage model. At the target point locations, climate model grid data can be downscaled using a variety of dynamic and statistical downscaling techniques. The application of the delta-change factor (also known as perturbation factors) is a straightforward technique for scaling up gridded climate forecasts to station scale. Delta-change variables have been used to design storm depth and precipitation time series.

This approach utilizes climate elements, often referred to as delta-change factors, to modify the model input based on historical observations or hypothetical design storms. To accommodate this disturbance, adjustments to the quantity of rainstorm events and the probability distribution of their intensities are necessary. Time resolution plays a critical role in flood analysis, making it imperative to convert daily rainfall data into hourly rainfall data. The daily time scale can be sub-divided into hourly units through various methodologies. Numerous stochastic downscaling approaches have been developed to further re-fine the RCM output temporally and transfer it to the spatial point scale. The combination of these interconnected models allows for the assessment of the impact of climate variability on the performance of sewer systems, specifically in relation to flooding events. Due to their limited capacity to reproduce extreme events, alternative downscaling strategies, such as weather typing or regression-based methods, have been found to be inadequate for this specific application.

2.2.3 Statistical Downscaling

The imprecise resolution and variability in precipitation outcomes from climate models necessitate a statistical model. This model cor-relates larger-scale atmospheric conditions (the *predictor* variables) with finer-scale rainfall patterns (the *predictand* variable), consider-ing spatial and temporal aspects. The model, which incorporates bias correction and statistical downscaling techniques, relies on historical data. It assumes that transferring information from predictors to pre-dictands will not significantly alter outcomes due to climatic variations. To generate data comparable to historical rainfall patterns, statistical downscaling is employed. This technique scales down climate model outputs, both spatially and temporally, to match the scale needed for urban hydrological impact modeling. The downscaling process further refines the data to accurately represent point rainfall. Existing statisti-cal downscaling techniques fall into three categories: empirical trans-fer function-based methods, resampling techniques-based methods, and conditional probability or stochastic modeling-based methods.

2.2.4 Methods Using Empirical Transfer Functions

Empirical transfer function-based techniques utilize the empirical relations or transfer functions between the precipitation predictand and its predictors. The statistical downscaling method proposed by Wilby et al. (2002), based on regression analysis, is well-recognized. Variables such as mean sea-level pressure, geopotential height, zonal wind speed and direction, specific or relative humidity, surface upward latent heat flux, temperature, dewpoint temperature, and dewpoint temperature depression have shown a strong correlation with small-scale precipitation at a daily or sub-daily level. This correlation suggests a relationship between these variables and the atmosphere's water vapor saturation level. Various geographical factors, including elevation, proximity to the coastline (diffusive continentality), advective continentality, and topographical slope, are believed to influence the vertical movement of the incoming air mass, potentially resulting in cooling and precipitation due to mountains.

Transfer functions such as generalized linear models, equations derived from rainfall time scaling principles, and artificial neural networks have been explored as regression relations. In urban drainage effect models that use continuous time series simulation and postprocessing of simulation results, it is standard to downscale the values in each time step to generate a rainfall time series. This downscaled time series is typically used in impact analysis of sewer overflows on receiving rivers. One approach involves preprocessing the time series data for both the predictor and predictand variables to derive relevant statistics, such as empirical frequency distributions or calibrated probability distributions, at specific time and space scales. Transfer functions can then be established between these statistics or distributions. Using artificial storms for specific storm frequencies or return periods can be beneficial in impact studies related to sewer surcharging or floods.

An alternative approach integrates a commonly used statistical downscaling model for spatial downscaling. This model links large-scale climate variables from GCM simulations with daily extreme precipitation events at a specific local site. Additionally, generalized extreme value (GEV) distribution is used for temporal downscaling to describe relationships between daily and sub-daily extreme precipitation occurrences. T. Nguyen and V. Nguyen (2018) used GCM climate simulations, National Centers for Environmental Prediction reanalysis data, and daily and sub-daily rainfall data from various rain gauges in Quebec, Canada, to validate this spatial-temporal downscaling approach. Adequate agreement with observed daily values at the site can be achieved by applying a bias-correction adjustment, based on a second-order polynomial function, to the annual maximum daily rainfall downscaled from the GCM. Following the collection of bias-corrected downscaled yearly maximum daily rainfalls at a specific location, T. Nguyen and V. Nguyen used a GEV distribution to further downscale sub-daily maximum rainfall intensities.

Probability distributions for rainfall intensities at sub-daily time scales (such as 5-, 15-, 30-minute, or hourly intervals) can be accurately determined by leveraging the distribution of daily rainfall intensities. This is achieved by applying the concept of scale invariance, where the moments of rainfall distribution (specifically, GEV distribution) are influenced by the time scale and its scaling properties. Introducing a dependency of the transfer function on RCM process variables, which include weather conditions such as cloud cover and precipitation type, could facilitate further advancements. To determine the wet proportion associated with different types of precipitation, researchers analyzed 30-minute measurements of various factors related to cloud cover. The average precipitation for the grid box was then converted into a local intensity, with an associated occurrence probability at each specific grid box point. It is important to note that the projected local intensity, type of precipitation, and cloud cover predicted by the RCM are subject to uncertainty. However, an evaluation conducted in Stockholm, Sweden, demonstrated that the technique aligns well with both empirical observations and theoretical considerations.

2.2.5 Strategies for Weather Typing or Resampling

Resampling, also known as weather typing, is a key component in some statistical downscaling methodologies. These methodologies use historical time series data of the region's rainfall predictand and coarse-scale climate predictor factors to obtain downscaled projected precipitation values. To determine downscaled future rainfall, the historical series of climatic variables are examined for each future event, such as a specific day, in the climate model output. This involves searching for a similar circumstance or analog event in the historical data. The small-scale precipitation observation corresponding to that particular occurrence is then considered as the downscaled future rainfall. Pressure fields derived from climate models are often used as predictive variables. Various types of weather are classified based on pressure fields using a categorization scheme.

2.2.6 Stochastic Models for Rainfall

The third category of statistical downscaling can be seen as an extension of stochastic hydrology. Stochastic rainfall models, mathematical constructs used for simulating and predicting rainfall patterns, use random variables and probability distributions to account for the inherent uncertainty and variability in rainfall data. These distributions are conditioned on the coarse-scale climatic predictor. The parameters of the stochastic model are derived from statistical analysis of time series data and can be adjusted based on climate model simulation results. Stochastic rainfall models are often referred to as *weather generators*.

Stochastic rainfall models employ a two-step process. First, the rainfall generator captures the structure of hourly storms. Then, the hourly rainfalls are refined to finer scales using a multi-scaling-based disaggregation approach. When transitioning from RCMs to urban catchment scales, delta-change techniques are commonly used. They involve identifying characteristics or variables assumed to remain consistent across different scales.

2.2.7 Variability in Hydrology

Climate change is expected to influence both the variability and average hydrology. Regions with minimal annual runoff variations may experience more frequent unusually low or high flow levels. Arnell (2003) assessed the potential impact of climate change on hydrological variability in six UK basins. The study revealed a slight increase in average monthly flow and a decrease in low flow levels by up to 40% by the 2080s. Additionally, there was an observed increase in year-to-year hydrological pattern fluctuations. The intensification of flooding, linked to climate change, is a significant concern worldwide, particularly in countries at lower elevations in tropical and humid mid-latitude zones. Major floods in significant river basins worldwide have increased. Kleinen and Petschel-Held (2007) found that approximately 20% of the global population lives in river basins that may experience more frequent flooding due to climate change. This was determined by applying statistically downscaled climate change projections to a water balance equation. Palmer et al. (2002) projected a five-fold increase in monsoons in Asia and heavy winter rainfall events in the UK. Lehner et al. (2006) also projected an increase in flood frequency in their continental-scale modeling analysis.

Kundzewicz et al. (2005) suggested that anthropogenic climate change might have contributed to previous large floods in central Europe and could influence future ones. Kay et al. (2006) observed increases in flood frequency and amplitude in most of their 15 UK study basins using a conceptual model driven by high-resolution RCM outputs into the 2080s. Despite a decline in mean annual runoff, Evans and Schreider (2002) observed an increase in flood size in six Australian basins using a conceptual hydrological model driven by stochastic weather generator output. Due to changes in temperature and precipitation, Mote et al. (2003) predicted an increase in winter floods in smaller, rainfall-dominated, transitory basins in the Pacific Northwest.

2.3 UNCERTAINTIES IN THE ANALYSIS OF CLIMATE CHANGE EFFECTS

Uncertainty in climate change impact studies often stems from various sources, including climate model projections, hydrology models, and data downscaling methods. The primary sources of uncertainty in climate model projections are internal variability, external forces, and model response. Uncertainties associated with climate variability and its prediction can be categorized into four groups: emissions scenario uncertainty, GCM uncertainty, downscaling uncertainty, and internal climate variability uncertainty. The design of drainage systems, which depends on extremely high amounts of precipitation occurring over a short period, can introduce sampling error. The uncertainty surrounding future urban development further complicates drainage system design. Consequently, higher population densities could alter the runoff coefficient in the future.

2.3.1 Climate Variability and Prediction Uncertainties

Models serve as the primary tool for projecting future climate change, enabling informed decisions about resilience initiatives related to climate change-induced drainage infrastructure. Therefore, it is crucial to characterize and quantify the uncertainty in climate change projections. These projections should be interpreted cautiously until the models can accurately reproduce historical temperature and precipitation ranges. In general, future climate change variability and prediction are uncertain due to model response, internal variability, and external forces. Model uncertainty arises from variations in physical and numerical formulations, leading to divergent responses among different models when subjected to identical external forces. Internal variability refers to the natural fluctuations within the climate system that occur in the absence of any external influences or forces. This encompasses a range of processes, including those related to the atmosphere, oceans, and their interaction. Uncertainty arises from an incomplete understanding of external factors that impact the climate system, such as future GHG emissions, stratospheric ozone levels, and changes in land use. This section provides a succinct overview of downscaling sampling and model uncertainties in hydrology.

2.3.1.1 Downscaling Sampling Uncertainty

The stochastic downscaling technique can generate a precipitation series of any length. Research shows that using a lengthy stochastic precipitation series instead of short observed samples can reduce sampling error in sewer system design:

1. The stochastic downscaling model can replicate all statistical features of precipitation with high precision, particularly those associated with extreme events.

2. The precipitation process exhibits annual consistency, assuming that the stochastic downscaling model does not explicitly account for inter-annual climate variability.

3. Sampling error is minimal when obtaining statistics from the observed precipitation series that was used to calibrate the stochastic downscaling model.

2.3.1.2 Model Uncertainties Related to Hydrology

Choosing a hydrological model in a climate impact assessment introduces additional uncertainty. Hydrological models can replicate runoff at various spatial and temporal scales due to variations in their parameters and underlying assumptions. To account for the socioeconomic components of the hydrological system where additional uncertainty may exist, practitioners can use results from hydrological models when preparing water resource management models. These include models of water demand, dam and reservoir storage, or policies promoting efficiency and conservation. Assumptions are necessary at every stage of the modeling chain; errors are inevitable and increase uncertainty in the modeling process.

2.4 FACTORS ACCELERATING EFFECTS ON URBAN DRAINAGE SYSTEMS

Research using climate change projections suggests that heavy precipitation events are expected to increase in both frequency and intensity. Numerous studies have found that areas with more impervious surfaces and higher rainfall experience more flash floods, flooding, and high peak flows. To manage stormwater effectively in the face of climate change, it is essential to establish a scientific method for comparing the impacts of global warming and urbanization on local precipitation. Local weather and climate are considered throughout the planning stages of stormwater management. However, the amount of stormwater runoff that must be managed can be significantly influenced by climatic changes, such as the number, frequency, and intensity of rain events, as well as land development.

Certain areas of developing countries may be particularly vulnerable to stormwater-related floods due to the combined effects of climate change and land-use change, while other parts of the world may remain mostly unaffected. Past anthropocentric approaches to stormwater management have had severe ecological consequences. The recent boom in suburbanization has contributed to both climate change and the loss of forested and agricultural land. Local hydrological cycles have been affected by increased surface runoff and decreased base flow, interflow, and depression storage. Several studies have linked impermeable surfaces to a 50% increase in surface water runoff and a 50% reduction in deep water penetration.

2.5 CLIMATE CHANGE AND FLOODING DUE TO EXTREME WEATHER EVENTS OR THE "NEW NORMAL"

The ongoing impact of global climate change has led to a noticeable increase in the occurrence of floods due to intensified variability in weather patterns. The modification of land cover, including the reduction of vegetation and the impact of climate change, exacerbates flood susceptibility. Extreme floods can occur due to various factors, including intense precipitation, prolonged duration, frequent precipitation, or a combination of these elements. An increasing number of coastal and rural villages are becoming submerged in water.

The recognition of climate change impacts, such as severe weather events and rising sea levels, is becoming increasingly prevalent. Floods occur when inland bodies of water, such as rivers and streams, tidal waters, or an excessive accumulation of water due to factors like intense precipitation or the failure of dams or levees, become inundated. Table 2.1 provides a comprehensive overview of various types of flooding, along with their corresponding descriptions.

Table 2.1   Types of main floods caused by climate change

1.	River inundation	Typically, arid land experiences inundation when a river or stream exceeds its regular boundaries. River flooding occurs most frequently during the late winter and early spring seasons. Potential causes include ice jams, heavy precipitation, or rapid snowmelt.
2.	Coastal inundation	This phenomenon occurs when elevated or increasing tides, storm surges, or coastal winds generated by a weather event (e.g., a hurricane) propel a significant volume of water from the ocean onto the adjacent land, resulting in the submergence or inundation of areas that are typically dry along the coast.
3.	Flash floods	The primary cause of these rapidly escalating floods is intense precipitation that typically occurs within a short duration, usually lasting no more than six hours. Flash floods can occur in various areas, although areas with insufficient drainage and lower elevations are more prone to their occurrence. Flash floods can occur as a consequence of dam or levee failures, abrupt water surges caused by debris or ice blockages, or other incidents that combine the inherent hazards of a flood with rapidity and unpredictability. They are responsible for the highest number of fatalities resulting from flooding incidents.
4.	Urban flooding	Urban flooding arises when the amount of rainfall surpasses the capability of drainage systems, such as storm sewers, to handle it. This can also occur when the expansion of roads, parking lots, and other impermeable surfaces in urban areas prevents water from infiltrating into the ground. The development and urbanization of an area, including the increase in pavement and other impermeable surfaces, typically leads to a reduction in the time it takes for water to flow through the hydrologic system.

2.5.1 How Global Warming Affects Precipitation

Changes in rainfall and other types of precipitation are among the most important aspects in evaluating the overall impact of climate change. Increased evaporation due to rising temperatures causes more severe precipitation. Average global precipitation has increased along with rising average global temperatures. Since the 1950s, extreme precipitation events have increased in frequency and caused heavier rain in many parts of the world. The Midwest and Northeast sections of the United States have seen the largest increases in heavy precipitation occurrences. These tendencies will persist as the world continues to warm. More water vapor can be held in warmer air—the air's capacity for water vapor increases by around 7% for every degree of heat. Higher intense precipitation episodes can result from an atmosphere with more moisture, and this is exactly what has been observed.

2.5.2 Rainfall Analysis

Outcomes of rainfall analyses significantly influence the design of urban drainage systems. The initial hydrologic study phase involves evaluating and predicting the expected precipitation levels within the designated study period. The following factors are significant in the context of urban drainage system design:

- *Rainfall duration*: how long a storm lasts
- *Frequency*: how often rainfall occurs at a particular amount, intensity, and duration
- *Rainfall depth to intensity*: rainfall depth divided by duration
- *Rainfall distribution*: the cumulative temporal and spatial distributions of rainfall across an area during a storm

2.6 AVERAGE RAINFALL CALCULATIONS

Rainfall depths observed at specific locations within the watershed are used to estimate the variation in rainfall depth over an area. Without a high density of rain gauges, it is typically impossible to accurately estimate the rainfall pattern and average values of rainfall depths. Rainfall recording has several levels of precision. The most precise recordings come from first-order weather service stations, which generate a continuous time-depth sequence that is often converted to an hourly sequence. The hydrologic network's recording-gauge data, provided for clock-hour intervals, rank second. These are transformed to produce hourly data. Nonrecording gauges, which measure daily rainfall depths, are also available.

Most of the time, the average depth of precipitation over the watershed is calculated using information gathered from specific gauging stations dispersed across a region. For a particular rainfall event, the average rainfall depth over a watershed can be calculated in three ways:

1. *Gauging station method*: Also known as an arithmetic average, this method is used in data analysis to determine the average value of a set of numerical measurements. This method produces accurate estimates when the terrain is level, the gauges are evenly spaced, and the individual gauge catches deviate minimally from the average. This approach involves collecting data on annual precipitation from multiple stations in the designated region. It offers a simple method for calculating the average precipitation in a specific catchment area. This is achieved by aggregating the recorded annual precipitation from all stations and dividing it by the total number of stations, as in Equation 2.1:

$$P_{ave} = \frac{1}{n}[p_1 + p_2 + p_3 \dots \dots \dots p_n] = \frac{1}{n}\left(\sum_{i=1}^{n} p_i\right), \quad (2.1)$$

   where N is the total number of stations and P_i is the mean annual precipitation at the ith station.

2. *Thiessen polygon method*: This method involves delineating polygon lines on a map by connecting neighboring rainfall gauge locations, which form equilateral triangles. The polygons around each station are created by the perpendicular bisectors of these lines. The area of each polygon is calculated through planimetry and expressed as a percentage of the total area. Each gauge is assigned a specific weight. The weighted average rainfall for the entire area is computed by multiplying the precipitation recorded at each station by its corresponding area percentage and summing these values. The results from this method are considered more reliable than those from

basic arithmetic averaging. Figure 2.1 illustrates the geometrical construction of this method and the area-based weighting applied to the rainfall value at each station.

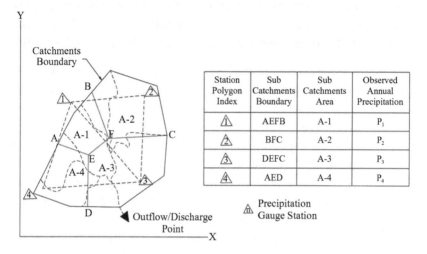

Station Polygon Index	Sub Catchments Boundary	Sub Catchments Area	Observed Annual Precipitation
⚠1	AEFB	A-1	P_1
⚠2	BFC	A-2	P_2
⚠3	DEFC	A-3	P_3
⚠4	AED	A-4	P_4

⚠ Precipitation Gauge Station

Figure 2.1   Geometrical construction of the Thiessen polygon method

In this method, the practitioner selects a consistent scale for the X and Y axes and draws the catchment area's boundary and the location of each station. Polygons are created by connecting adjacent stations and drawing perpendicular bisectors between them. The area of each polygon is determined by the sum of its box counts. The next step involves calculating the product of p_1A_1 and summing all the products. The average precipitation can then be calculated as shown in Equation 2.2:

$$P_{ave} = \frac{p_1A_1 + p_2A_2 + p_3A_3 \dots \dots p_nA_n}{A_1 + A_2 + A_3 \dots \dots \dots + A_n}, \quad (2.2)$$

where N is the number of polygons in the catchment area, P_n is the observed annual rainfall for the ith polygon, and A_n is the area of the ith polygon.

3. *Isohyetal method*: This method involves plotting the locations and values of stations on a map and delineating contours

representing equal precipitation levels (isohyets), as shown in Figure 2.2. The average precipitation over an area is computed by multiplying the average precipitation between successive isohyets by the watershed area located between these isohyets, summing these products, and dividing the sum by the total area.

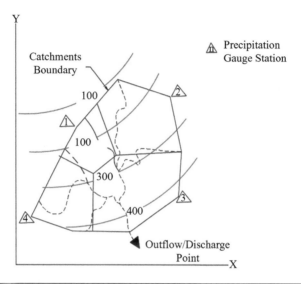

Figure 2.2   Representative contour map area of the isohyetal method

In this method, the first step is to select a consistent scale for the X and Y axes. The second step involves drawing the catchment area's boundary and the location of each station. The third step is to determine the rainfall amount at each station and the appropriate contour interval and number of isohyets. The fourth and fifth steps involve drawing isohyets between stations using linear interpolation and calculating the distance between two consecutive isohyets. The product $P_i A_i$ is then calculated. The average precipitation can be calculated using Equation 2.3:

$$\bar{p} = \frac{\sum_{i=1}^{n} A_i \frac{(p_i + p_{i+1})}{2}}{(A_1 + A_2 + A_3 + \ldots A_n)}. \tag{2.3}$$

2.6.1 Estimating Missing Rainfall Data

Brief lapses in precipitation recordings at some stations may occur due to the observer's absence or equipment malfunction. It is typically necessary to estimate this gap in the record. The estimation of missing data can be achieved using the methods outlined in the following sections.

2.6.1.1 Arithmetic Mean or Local Mean Method

This method uses simultaneous rainfall data from three nearby stations that are evenly distributed around the station with the missing records. The estimated value of the missing data is obtained by taking a simple arithmetic average of the rainfall at the three selected stations. This technique can be used to compute missing monthly and annual rainfall values. This method may only be used when the yearly precipitation at each station is within 10% of the station without records, as shown in Equation 2.4:

$$P_x = \frac{1}{N}[p_1 + p_2 + p_3 \ldots \ldots \ldots p_n] = \frac{1}{N}\sum_{i=1}^{N} p_i, \qquad (2.4)$$

where p_x is the average annual precipitation at X station, p_i is the annual precipitation recorded at the ith rain gauge station in the catchment, and N is the total number of rain gauges.

2.6.1.2 Normal Ratio Method

The normal ratio (NR) method involves assigning weights to rainfall data based on the ratios of normal annual rainfall values. This is applicable when the normal annual rainfall of a selected station constitutes 10% or more of the station with missing records, and the simple average method is not suitable; refer to Equation 2.5:

$$P_x = \frac{N_x}{m}\left[\frac{p_1}{N_1} + \frac{p_2}{N_2} + \frac{p_3}{N_3} \ldots \ldots \ldots \frac{p_m}{N_m}\right]. \qquad (2.5)$$

where p_x is the missing annual precipitation at x station, $p_1, p_2, \ldots, p_m$ are the annual precipitation at $1, 2 \ldots, m$ stations, N_x is the normal annual precipitation at the stations around x, and $N_1, N_2, \ldots, N_m$ are the normal annual precipitation at the $1, 2, \ldots, m$ stations.

2.6.1.3 Modified NR Method

The modified NR method can account for the influence of distance when estimating missing precipitation data; refer to Equation 2.6:

$$P_x = \frac{\sum_{i=1}^{n} D_i^{1/b} \left(\frac{\overline{P}_x}{\overline{p}_i} \right)}{\sum_{i=1}^{n} D_i^{1/b}}, \tag{2.6}$$

where P_x is normal rainfall, D_i is the distance between the index station i and the gauge station with missing data or ungauged station, n is the number of index stations, and b is the constant by which the distance is weighted (normally 1.5–2.0; commonly using $D^{0.5}$).

2.6.1.4 Inverse Distance Method

Among the methods discussed, the inverse distance method is recommended as the most accurate. The estimated rainfall at a location depends on the rainfall measured at surrounding index stations and the distance from the ungauged location to each index station. Rainfall P_x at station x is calculated using Equation 2.7:

$$P_x = \frac{\sum_{i=1}^{N} \frac{1}{d^2} p_i}{\sum_{i=1}^{N} \frac{1}{d^2}}, \tag{2.7}$$

where P_x is the estimate of rainfall for the ungauged station, P_i is rainfall values of rain gauges used for estimation, D_i is the distance from each location of the point being estimated, and N is the number of surrounding stations. Moreover, $d = 2$ is commonly used. Given that the weighting in the inverse distance approach is dependent on distance, this method is not suitable for use in hilly areas.

2.6.1.5 Linear Programming Method

The linear programming (LP) method involves selecting a base station and multiple adjacent index stations to determine the optimal weighting factor. This is achieved by minimizing the difference between

observed and computed rainfall at the base station across various rainfall events. The method computes the optimal weighting factors for
the base station and its associated index stations, aiming to minimize
the total sum of deviations for a set of K events; refer to Equations
2.8 and 2.9:

$$\sum_{j=1}^{k}(U_j + V_j),\qquad(2.8)$$

subjected to

$$\sum_{i=1}^{n}(a_i r_{ij} - U_j + V_j) = r_{bj}\ (j = 1,2, ... K),\qquad(2.9)$$

$\sum_{i=1}^{n} a_i = 1.0$ (sum of weights is 1),

$a_i \geq 0, \qquad U_j \geq 0, \qquad V_j \geq 0,$

where i is the index station, j represents the index for rainfall events,
and b represents the observed rainfall at base station b for event j.
$\sum_{i=1}^{n}(a_i r_{ij})$ calculates the amount of rainfall at the base station for
event j.

For any event, $CR - OR = \delta$, where CR is the computed rain, OR is
the observed rain, and δ is the deviation. The outcome can be positive
or negative without any restrictions on its sign. In LP, these variables
are substituted with the difference between two nonnegative variables.

2.6.2 Depth-Area-Duration Relation

Depth-area-duration (DAD) describes the relationship between the
area distribution of a storm and its duration. A DAD analysis assesses

the maximum rainfall quantities across different durations and areas during a storm. Analyzing and processing raw rainfall records in the region can yield valuable information in the form of curves or statistical values, which is useful for water resource development projects. It is crucial to analyze the temporal and spatial patterns of storm precipitation to address various hydrologic issues. The average depth of rainfall decreases exponentially with increasing area for a given rainfall duration; refer to Equation 2.10:

$$\bar{P} = P_0 e^{-(KA^n)}, \tag{2.10}$$

where P is the average depth, measured in centimeters, across a given area (A) in km^2 and P_o is the maximum recorded rainfall in centimeters at the center of the storm; k and n are constants that remain fixed within a specific region.

Preparing DAD curves requires considerable computational effort and depends on the availability of region-specific meteorological and topographical information. Generally, these are the steps that are followed:

1. Analyze the historical precipitation data for the geographical area where the catchment area that is under consideration is located, considering records from places with similar meteorological conditions.
2. Compile a detailed list of the most severe storms, including their dates of occurrence and durations.
3. Generate isohyetal maps and compute the corresponding rainfall values for each isohyet within the designated area with the severe storms that were analyzed in Step 2.
4. Use a graph to illustrate the correlation between area and rainfall amounts for different time periods, such as one-, two-, or three-day rainfall.

2.6.2.1 Use of DAD Curves

The depth-area-duration curve is a useful tool for analyzing storm precipitation in relation to time and area. It allows us to determine the maximum amounts of precipitation for different durations and areas. Figure 2.3 displays the precipitation depths for the proposed development catchment for durations of one, two, and six hours.

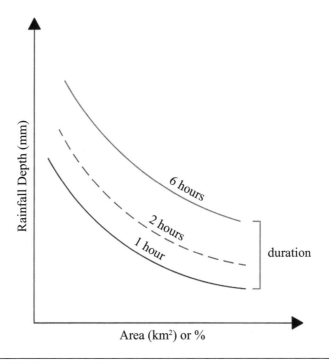

Figure 2.3   Depth-area-duration curve

2.7 INCORPORATING EXTREME WEATHER IN URBAN DRAINAGE SYSTEMS

A critical initial task in the diagnostic framework is discerning the types and potential magnitudes of climatic changes that could impact urban drainage systems. This information can be utilized to enhance the resilience of urban drainage systems against changes that might threaten system functionality and infrastructure. Climate change planning must consider that, while experts and authorities recognize significant ongoing climatic changes and expect these changes to continue, no study has completely clarified their exact nature, especially at the local level. This uncertainty poses a substantial challenge for those managing urban drainage infrastructure systems potentially affected by climate change. Therefore, it is fundamental to consider the uncertainties associated with anticipated climatic changes when making investment or operational decisions.

The frequency and intensity of extreme rainfall events have fluctuated over recent decades, a trend attributed to changes in weather and climate. Urbanization could exacerbate these variations in intense precipitation due to the urban heat island effect. Understanding the factors contributing to these changes is crucial for estimating the societal, economic, and environmental impacts of extreme rainfall. The frequency and volume of stormwater flows would increase, necessitating upgrades to drainage infrastructure. This situation presents a significant challenge to urban drainage management and related fields. Historically, drainage systems were built using what is often termed *gray infrastructure*, which generally lacks the adaptability needed to handle intense precipitation resulting from extreme weather, the so-called "New Normal."

The primary purpose of drainage system networks is to collect stormwater runoff from designated precipitation events and direct it to wastewater treatment facilities or downstream within a municipal separate storm sewer (MS4) system. The process of capturing and transferring stormwater runoff adheres to established guidelines and standards for drainage system design. However, their primary function does not include mitigating flooding resulting from severe weather events.

The traditional approach to designing urban drainage system networks typically involves determining a return period, denoted as T, representing the frequency of an event's occurrence. It is assumed that the event will occur with a probability of $1/T$ per year. However, extreme rainfall events are seldom considered in drainage system design, except for scenarios where the drainage system is overwhelmed.

Designing drainage system networks to effectively handle severe rainfall events could result in the development of drainage infrastructure that may not be cost effective. To manage the drainage system comprehensively, it is necessary to consider a wider range of precipitation events. In the context of collector networks, design events typically occur within a time frame ranging from 1 to 10 years. However, exceedance events are characterized by a longer return period, typically from 50 to 100 years. Extreme events are even less frequent, with a return period extending from 100 to 1,000 years. Determining an optimal threshold value remains a significant challenge.

Contrary to the characteristics of design events, managing exceedance flows and flooding for a specific frequency of occurrence or amount of rainfall is a complex task. One factor to consider is the feasibility of implementing a stringent threshold or criterion, especially when it needs to be applied to an entire city, drainage area, or catchment. This is particularly true for urban areas where the regulation of exceedance flows has not been effectively implemented. Managing exceedance in these areas may incur significant adaptation costs when employing a rigorous threshold or criterion. For instance, modifications to streets or building components may be necessary to meet the recommended threshold or criterion. While establishing management standards for construction in low-risk or safe locations for new developments is common practice, the presence of extensive catchments can lead to hazardous flooding situations. This necessitates the implementation of structural measures such as protective channels or embankments that comply with more stringent design criteria, such as those based on a 100-year return period. Despite the challenges in effectively regulating excessive flows and floodwaters, it is crucial to prioritize infrastructure and urban planning adaptations to minimize economic losses, especially during catastrophic events. Several methodologies for understanding the importance of and assessing adaptation measures require an analysis of the impacts of precipitation or other climate-related phenomena.

Risk assessment is a commonly employed methodology within the realm of urban drainage systems to evaluate and mitigate the potential consequences of severe weather events effectively. This process involves the methodical assessment and execution of adaptive strategies aimed at mitigating risks to both ecosystems and human well-being. Conducting a risk assessment involves identifying potential risks, exposures, and vulnerabilities. The spatial distribution of social groups and properties susceptible to impacts is determined by exposure and vulnerabilities. Hazards are commonly defined by their return period, which pertains to the frequency of occurrence of external loadings. Uncertainty is a common occurrence in risk assessments due to the difficulties associated with assigning probabilities to socioeconomic and climate change scenarios, evaluating damages, and calculating the costs of adaptation activities.

2.7.1 Extreme Event Probabilities

The assessment of flood risk, along with the development and implementation of flood mitigation strategies, relies significantly on the application of probabilistic models for intense precipitation. Gumbel distribution has been widely accepted as the primary model for intense precipitation phenomena. The use of Gumbel distribution, which has an exponential tail in the underlying distribution, is supported by both theoretical and empirical findings. However, the suitability of this distribution has recently been questioned due to both theoretical and empirical considerations.

Recent theoretical investigations suggest that *extreme value type II* distribution may be a more appropriate substitute for Gumbel distribution. These analyses involve comparing actual and asymptotic extreme value distributions and applying the principle of maximum entropy. Furthermore, several empirical studies have been conducted using extensive rainfall data to support these recent theoretical findings. Additionally, empirical analyses have shown that Gumbel distribution tends to underestimate the most extreme rainfall levels. However, it is important to note that this particular distribution provides excellent predictions for shorter return periods, specifically those up to 10 years.

It is worth emphasizing that Gumbel distribution may be considered an appropriate model for situations involving a limited number of years of measurements, particularly when using subsets of extensive data sets.

2.7.2 Analysis of Rainfall Data Using Statistical Methods

Precipitation is a critical factor in many hydraulic engineering applications, including the planning and design of hydraulic structures such as bridges, culverts, canals, and storm sewage drainage systems. Accurately determining the relevant input value for the design and construction of engineering structures requires a comprehensive statistical analysis of each specific region. Caution is necessary when performing a frequency analysis on precipitation data since the shape of flood-frequency distributions may vary depending on the equations used in the analyses. Therefore, it is imperative for practitioners to have a thorough understanding of the terminology used in frequency analysis. The most commonly used analytical methods are the normal, log-normal, Gumbel, and log Pearson type III distribution methods.

2.7.3 Expected Rainfall Depths for a Given Probability

Accurate assessments of precipitation depths or intensities expected with a specific level of probability within a designated time frame (ranging from 1 to 24 hours, daily, weekly, monthly, or yearly) are crucial for the strategic planning and implementation of urban drainage initiatives. The term *probability* refers to the likelihood of exceeding a specific threshold and represents the chance that the actual amount of rainfall within a specified time frame will be *equal to* or *greater than* the estimated rainfall depth. *Rainfall depth* describes the amount of precipitation expected or that may be exceeded during a specific period, based on a given probability. The *minimum reliable rainfall threshold* refers to the lowest amount of rainfall that can be considered reliable within a specified time frame.

2.7.4 Probability of Exceedance

The probability of exceedance refers to the likelihood of a specific event or value surpassing a predetermined threshold. This probability can be expressed as a percentage from 0% to 100%, or as a fraction between zero and one. The projected amount of rainfall that could occur or be exceeded in a given year within a specific time period can be quantified as a numerical value representing the number of years within the defined time frame.

2.7.5 Recurrence Interval

Recurrence intervals are used to evaluate the average period between rainfall occurrences of similar or larger size. The variability of rainfall patterns is influenced by several factors, including the duration and intensity of precipitation events, and the geographical context. The assessment of the likelihood of a particular quantity of precipitation occurring during a specified year is commonly conducted through recurrence intervals, also known as return periods. The return period, a widely recognized unit of measurement, is often stated in years. It is derived by evaluating the likelihood of exceeding a storm event. The concept of *likelihood* refers to the possibility of a storm of a specific magnitude occurring or exceeding a storm during an interval of one year. Equation 2.11 is a mathematical tool that is used to determine the correlation between the recurrence interval and exceedance probability:

$$T = \frac{1}{P} \tag{2.11}$$

where T represents the return duration measured in years and P is the likelihood of exceedance. A 20% dependable rainfall ($PX = 0.20$) has a return period of $1/.2 = 5$ years, meaning that, on average, the rainfall in the first decade of January will exceed 23 mm in Tunis once every five years. A 50% dependable rainfall has a return period of two years, indicating that the rainfall depth is exceeded, on average, once every two years.

2.7.6 Probability of Exceedance for Design Purposes

The calculation of the probability of exceedance (P) or return period (T) for design purposes is influenced by several factors, including the potential damage caused by excessive rainfall, the acceptable level of risk, and the projected lifespan of the project. The determination of a design return period is not solely based on an economic evaluation comparing the costs and benefits of implementing drainage infrastructure. It also involves a comprehensive policy decision that considers factors such as land use and potential risks to public safety. A pragmatic approach is recommended. In specific instances, such as temporary river diversions, the criteria for determining the appropriate return period for design purposes may be overly stringent. Therefore, it is advisable to establish the design return period by relying on local expertise and conducting a comprehensive risk assessment. This assessment should consider various factors, including the project's duration, the specific seasons during which the project will be carried out, and any additional contingency measures that may be required. Table 2.2 can be a valuable resource when used in conjunction with applicable local regulations and practical expertise. It is essential to acknowledge that the suggested choice of return period may not always be suitable or attainable, especially when considering the implementation of new drainage systems or the upgrading of existing ones, particularly in low-lying areas or densely populated urban locations.

Table 2.2 Recommended return periods for drainage systems and projects

Type of drainage systems/project	Return period in years
Flood plain development	100
Urban drainage—Low risk (up to 100 ha)	5 to 10
Urban drainage—Medium risk (more than 100 ha)	25 to 50
Urban drainage—High risk (more than 1,000 ha)	50 to 100
Road drainage	25 to 50
Highway drainage	50 to 100
Bridge design—Piers	100 to 500
Levees—Medium risk	50 to 100
Levees—High risk	200 to 1,000
Principal spillways—Dams	25 to 100
Emergency spillways—Dams	100 to 10,000 (PMP)*
Freeboard hydrograph—Dams Class (c)	10,000 (PMP)*

* Probable maximum precipitation (PMP)
Note: adapted partly from "What are the return periods commonly used in design?" Dr. Victor Miguel Ponce, San Diego State University (https://ponce.sdsu.edu/return_period .html), and different worldwide city municipalities drainage design manuals.

2.7.6.1 Probability of Design Failure

The return period (T), design life (n), and probability of exceedance (P) are critical considerations when designing drainage system networks. It is important to clarify a common misconception: specifying a system for a T-year return period event does not mean the system's capacity will only be exceeded once every T years. The T-year return period is a statistical measure used to evaluate the probability of an extreme event occurring within a specified duration. The term *frequency* refers to the average rate at which an event of a specific magnitude is expected to occur. In reality, stochasticity and natural variability affect the occurrence of extreme events, and it is possible for multiple events of equivalent magnitude to occur in close succession or at extended intervals.

However, the design life of drainage infrastructure, represented by n years, refers to the projected duration during which the drainage system is expected to operate efficiently without significant problems or failures. This measure is used as a reference point for planning and constructing drainage systems, ensuring that the system is built to withstand projected potential risks within a designated time frame. Therefore, it is crucial to consider the probability (P) of a drainage system's capacity being exceeded at least once over its design lifespan. This consideration helps us understand the likelihood of encountering a design that fails to meet its intended performance, calculated using Equation 2.12:

$$P = 1 - \left(1 - \frac{1}{T}\right)^n \tag{2.12}$$

Consider proposed drainage infrastructure designed to function for 30 years, with a 45% likelihood of failure within this lifespan. To mitigate this risk, it is advisable to engineer infrastructure to withstand a 51-year recurrence interval or a 51-year peak flow. This approach will help address potential challenges and enhance the long-term resilience of this piece of infrastructure.

2.7.6.2 Probable Maximum Precipitation

Probable maximum precipitation (PMP) refers to the maximum depth of precipitation achievable within a specific area and duration, without exceeding known meteorological conditions. In simpler terms, PMP represents the upper limit of precipitation expected under the most extreme weather circumstances. Understanding PMP is crucial for various applications, including infrastructure design, water resource management, and flood risk assessment. By determining the PMP for a particular region, meteorologists and engineers can make informed decisions about the design and capacity of structures like dams, reservoirs, and drainage systems.

It is important to note that calculating PMP involves considering a range of meteorological factors, including atmospheric moisture content, wind patterns, and topographical features. These factors collectively influence the potential for precipitation in a given area. By analyzing historical weather data and using sophisticated mathematical models, meteorologists can estimate the maximum amount of precipitation that can occur under extreme conditions.

The PMP value serves as a benchmark for planning and designing infrastructure to withstand severe weather events. It provides assurance that the structures will withstand the most intense weather conditions. PMP is widely used in the planning and execution of extensive hydraulic infrastructure projects, particularly in large-scale dam construction. These projects often involve the design and implementation of critical components, such as spillways, which play a crucial role in managing water flow and preventing potential damage to the dam structure. The application of PMP in the design of large hydraulic structures, including spillways in large dams, underscores the importance of project management in the successful execution of complex engineering projects.

Variations in PMP are observed worldwide, with significant differences based on the climatic regions across the globe. Various methodologies are used for calculating PMP, including statistical methods and the examination of storm mechanisms that give rise to intense precipitation events. In engineering, it is common practice to use one's judgment to determine an appropriate value for a given situation. This process involves carefully considering various factors and making a decision based on one's expertise and experience in the field.

PMP refers to the highest amount of rainfall that can occur within a specific time period at a rain gauge station or a basin. It represents the upper limit of rainfall intensity that is physically achievable in a given location. The concept discussed here pertains to the precipitation level that would result in a flood within a basin while ensuring that there is no possibility of surpassing the predetermined threshold.

In hydrology, PMP can be estimated by using a statistical approach. This estimation is given by Equation 2.13:

$$PMP = k\delta, \tag{2.13}$$

where k and δ represent certain parameters. PMP is the mean of the annual maximum rainfall series, representing the average value of the highest recorded rainfall in a given year. The parameter k, known as the frequency factor, depends on various factors. These include the statistical distribution of the rainfall series, the number of years of record available, and the desired return period. The return period refers to the average time interval between occurrences of a rainfall event of a certain magnitude. Finally, the parameter δ represents the standard deviation of the rainfall series. It quantifies the variability or spread of the annual maximum rainfall values around the mean. By combining these parameters in the equation, we can estimate the PMP. This estimation allows us to assess the maximum amount of rainfall that could potentially occur within a specific region or catchment area. The value of k ranges from zero to 15.

2.8 PLOTTING POSITION

Frequency analysis, a fundamental technique, can be approached from two perspectives: empirical and analytical. Each offers unique methodologies for data interpretation, enabling researchers to discern patterns and trends within datasets. Understanding these approaches allows for the effective use of frequency analysis, providing valuable information and accurate conclusions. This is particularly useful in the initial design phase of engineering projects that are focused on flood control and drainage systems.

2.8.1 Empirical Method

Frequency analysis assesses the probability of an event occurring within a specified time frame. Consider an event, such as rainfall, with the goal being to determine the probability of this event reaching or exceeding a certain magnitude, denoted as X. This probability is quantified using p. The Weibull formula provides the return period, or recurrence interval, associated with a given probability p; refer to Equation 2.14:

$$p = \frac{m}{N+1},\qquad(2.14)$$

where p is the exceedance probability of the event and m is the rank assigned to the data after arranging them in descending order of magnitude. Thus, the maximum value is $m = 1$, the second largest value is $m = 2$, and the lowest value is $m = N$, with N being the number of records.

The exceedance probability of the event is calculated using an empirical formula known as the plotting position. Numerous plotting position formulas have been developed and refined over time, serving as essential tools for accurate data representation and interpretation. Table 2.3 presents a comprehensive list of these formulas, which have demonstrated their utility in various applications (Subramanya 2006). The Weibull formula, often used as a plotting position, requires extensive historical data for a thorough investigation.

Table 2.3   Plotting position formulas

Method	P(probability)
California	$\dfrac{m}{N}$
Hazen	$\dfrac{m - 0.5}{N}$
Weibull formula	$p = \dfrac{m}{N + 1}$
Jenkinson's method	$\dfrac{m - 0.3}{N + 0.4}$
Gringoten	$\dfrac{m - 0.44}{N + 0.12}$

The procedure should include the following steps:

1. Calculate the exceedance probability for each data point for ranking and plotting position
2. Generate a probability plot for the data, selecting an appropriate distributional assumption

3. Evaluate the suitability of the chosen distribution, considering alternative distributions or modifying the data to fit the chosen distribution, if necessary
4. Determine realistic rainfall depths for specific probabilities or return periods using probability plots
5. Apply analytical techniques to incorporate the frequency factor effectively, yielding more accurate and refined results

2.9 THEORY OF EXTREME VALUE

Fisher and Tippet (1928) identified three limiting distributions for extreme value analysis (EVA), building upon the groundwork established by Fréchet (1927). Ludwig von Mises further developed extreme value theory (EVT) in 1936 by defining conditions for convergence, which Gnedenko formalized in 1943. Common distributional assumptions for modeling extreme rainfall data include the following:

1. Generalized extreme value (GEV) distribution
2. Log-normal distribution – 3 parameters (LN3)
3. The Pearson type III distribution (P3)
4. Generalized Pareto distribution (GP)
5. Gumbel distribution

2.9.1 Background

In educational contexts, the concept of a coin toss is often used to illustrate the principle of a binomial probability distribution. A coin toss, a simple yet fascinating method for decision making or determining outcomes, involves flipping a coin and observing the upward-facing side. This method is used in various scenarios, from casual games to significant events. In an ideal scenario, a coin has an equal 50% probability of landing on either heads or tails in a single trial. With multiple tosses, it is possible to gain insights into the probability of obtaining either outcome. This knowledge allows for the prediction of future trial outcomes, including the examination of the frequency ratio between heads and tails. However, this basic concept can be expanded to cover more complex cases, as shown by other probability distributions. The primary goal in drainage hydrology is to create a mathematical model that accurately represents rare or exceptional events with a low occurrence probability.

In drainage and hydrology, the term *extreme* typically refers to precipitation that surpasses the usual variability range within a specific geographic and temporal context. Modeling extreme weather phenomena presents challenges due to their sporadic occurrence, making the collection of accurate and reliable data difficult.

EVT is a statistical framework designed to address the inherent randomness observed in natural variability. Its aim is to characterize extreme events by quantifying their probability of occurrence. The frequency of events of different magnitudes can be described as a series of random variables with the same distribution. Let f represent a function that approximates the relationship between the event's magnitude, represented by X_N, and its occurrence probability. This relationship can be mathematically expressed as shown in Equation 2.15:

$$f = X_1, X_2, X_3, \dots . X_N.$$ (2.15)

The data derived from the resulting distribution can be used for trend analysis and assessing the probability of severe occurrences, which includes predicting the frequency and intensity of extreme weather precipitation. These distributions can also be used for simulations.

2.9.2 Generalized Extreme Value versus Generalized Pareto

GEV distribution is a prevalent method in EVA. It examines the distribution of block maxima, where a block refers to a specific time interval, such as a year. Depending on its shape parameter, GEV distribution can exhibit characteristics of Gumbel, Fréchet, or Weibull distributions. As an alternative approach, GP (generalized Pareto) focuses on analyzing values exceeding a predetermined threshold. The resulting distribution varies based on the shape parameter, leading to an exponential, Pareto, or beta distribution. These two methodologies are summarized in Table 2.4.

Table 2.4 Description of the two basic types of extreme value distributions; based on Pinheiro, M., and Grotjahn, R. (2015). An introduction to extreme value statistics. Tutorial at the University of California at Davis.

	Generalized Extreme Value (GEV)	Generalized Pareto (GP)
General Function (CDF)	For extreme value Z, $$G(Z) = exp\left[-\left\{1+\xi\left(\frac{Z-\mu}{\delta}\right)\right\}^{-1/\xi}_+\right]$$	For threshold excess χ, $$H(\chi) = 1 - \left[1+\xi\left(\frac{z-\mu}{\delta}\right)\right]^{-1/\xi}_+$$
Limit as $\xi \to 0$	Gumbel: $$G(Z) = exp\left[-exp\left\{-\left(\frac{Z-\mu}{\delta}\right)\right\}\right]$$	Exponential: $$H(\chi) = 1 - exp\left(\frac{z-\mu}{\delta}\right)$$
$\xi > 0$	Fréchet	Pareto
$\xi < 0$	Weibull	Beta
Description	Distribution function of standardized maxima (or minima) — block maxima/minima approach	Probability of exceeding pre-determined threshold peaks over threshold approach
Parameters	**Location** μ: position of the GEV mean	**Threshold** u: reference value for which GP excesses are calculated
Interpretation of results	Return level: value z_p that is expected to be exceeded on average once every 1/p periods, where 1 − p is the probability associated with the quantile. Find z_p such that G (z_p) = 1 − p	Return level: value x_m that is exceeded every m times. Begin by estimating ζu, the probability of exceeding the threshold. Then, x_m is $$x_m = \begin{cases} u + \frac{\sigma_u}{\xi}\left[(m\zeta_u)^\xi - 1\right] & \xi \neq 0 \\ u + \sigma_u ln((m\zeta_u)) & \xi = 0 \end{cases}$$

2.9.3 Stationarity versus Nonstationarity in the Field of Data Analysis

Stationarity, the consistency of statistical properties over time, is a crucial characteristic of data. In data analysis, *stationary* describes a state where key statistical measures, such as mean and variance, remain constant over a specific period. This constancy is vital in various data analysis methodologies because stationary data do not exhibit significant deviations. The assumption of stationarity is crucial in statistical modeling techniques and forecasting methods, enabling reliable predictions and meaningful insights. Evaluating a model's fit requires analyzing the temporal stability and consistency of the model's distribution, differentiating between stationarity (stable distribution over time) and nonstationarity (unstable distribution over time). Stationary models consistently represent variables, such as x, σ, and ξ, as time-invariant functions, using fixed constants as parameters.

Nonstationary models, lacking fixed constants as parameters, differ from stationary models. Understanding the distinction between stationary and nonstationary models is essential for interpreting data behavior and characteristics in various analytical contexts. Modeling nonstationary extremes typically involves a constant high threshold, denoted as x_0, with threshold exceedances modeled using the GP. To achieve a linear increase in the GP threshold or to incorporate seasonal cycles, the following equation is introduced into GP parameters, allowing for nonstationarity; refer to Equation 2.16:

$$x(t) = x_0 + x_1 t, \qquad (2.16)$$

where x_0 is the initial threshold value and x_1 denotes the rate of increase in the threshold value as time progresses. Similar adjustments can be made for the remaining variables.

2.10 CHAPTER SUMMARY

To thoroughly evaluate the resilience of urban drainage infrastructures against global climate change and local watershed responses, multi-scale modeling analysis is often essential. This approach considers the different scales at which these infrastructures operate, allowing for a comprehensive performance assessment. By considering the broader effects of global climate change and the specific responses of local watersheds, we can achieve a more precise evaluation of resilience.

This section explores climate model projections related to the occurrence of intense rainfall in future climates. The projections suggest an increased likelihood of intense rainfall events due to rising GHG levels. In the context of urban drainage systems, it is important to note that their design heavily relies on the statistical analysis of past data. Examining historical data and studying the outcomes of previous events provide valuable insights for making informed decisions aimed at ensuring our infrastructure can effectively manage rainfall. It is crucial to consider potential consequences, such as an increased frequency of flooding incidents, that may arise from an increase in the severity and frequency of extreme rainfall events.

When evaluating design criteria, it is essential to review and modify them to accommodate potential changes due to climate change. This revision involves considering three key factors: climate projections related to extreme rainfall in the region under investigation, the anticipated performance level or permissible risk level, and the projected lifespan of the infrastructure or system. Incorporating these factors allows for effective mitigation of climate change effects through suitable modification of the design criteria. The revised design criteria ensures that the service level consistently exceeds the chosen *acceptable* level throughout the predetermined lifespan of the infrastructure.

It is paramount to incorporate the definition of new design criteria into a comprehensive global adaptation strategy. The goal of this strategy is to integrate various measures to maintain a satisfactory service level in the long run. Determining the service level in light of uncertainties related to anticipated variations in heavy precipitation presents a significant challenge.

2.11 CHAPTER PROBLEMS

1. Define the following terms based on your own understanding:

   a. Climate
   b. Climate change
   c. Climate change adaptation
   d. Climate model
   e. Climate prediction
   f. Climate projection
   g. Climate risk
   h. Climate scenario
   i. Climate system
   j. Coastal erosion
   k. Extreme weather event
   l. Flood and flood mitigation
   m. Global warming
   n. General circulation models (GCMs)
   o. Hydrological cycle
   p. Regional climate models (RCMs)
   q. Resilience
   r. Risk

2. What are the key distinctions between climate change and global warming?

3. What is the function of climate models? Can these models be developed for regional climates?

4. Extreme weather events, including hurricanes, cyclones, and heavy rainfall, are natural phenomena with significant impacts on our planet. These events are defined by their intensity and duration. Investigate the intricate relationship between extreme weather events and rising sea levels.

5. Climate forecasting is essential for understanding and predicting future climate patterns, which is crucial for sectors such as agriculture, energy, and water resource management. However, it is necessary to acknowledge that climate forecasting is a complex process that depends on the use of sophisticated

models. These models aim to simulate and project future climate conditions based on various factors and variables. What are the critical considerations when utilizing current climate models for forecasting?

6. Discuss the potential effects of extreme weather events on urban drainage infrastructure, focusing specifically on its operation and maintenance.

7. Analyze the range of climate change adaptation strategies that municipal agencies could potentially adopt.

8. What role, if any, does asset management play in an agency's climate change adaptation efforts?

9. Considering climate-related risks is vital due to their various economic, environmental, and social impacts. By accounting for that climate-related risk, we can enhance our understanding and mitigation of potential climate change consequences. This understanding enables us to make informed decisions and take suitable actions to safeguard our environment, economy, and society. Climate-related risks include a broad spectrum of factors, such as extreme weather events, rising sea levels, changes in temperature and precipitation patterns, and ecosystem shifts. Discuss and analyze the practical application aspects, citing examples of completed or proposed projects.

10. In the context of risk assessment, can the outcomes of a risk assessment be represented without incorporating probabilities?

11. What is the probability of exceedance?

12. List and discuss the four basic types of rainfall models.

13. Discuss how cities are integrating extreme weather into their urban drainage systems.

14. Engineers recognize that climate change is a dynamic process, and they consider this when designing systems and structures. How does the field of engineering design tackle the challenges presented by an ever-changing climate?

15. Changing climate patterns and the increasing frequency and intensity of extreme weather events present substantial challenges and risks to construction projects. One of the primary concerns for the construction industry is the potential damage

caused by extreme weather events such as hurricanes, floods, and heat waves. These events can result in severe infrastructure damage, including the destruction of buildings, roads, and bridges. The increased occurrence of these events due to climate change can lead to significant financial losses for the construction industry. Discuss and analyze actual construction projects that have been affected by extreme weather events.

16. What is the difference between stationarity versus nonstationarity in the field of data analysis?

17. Table 2.5 provides data for the base station and the four surrounding stations. Using (*i*) the modified normal ratio (NR) method and (*ii*) the inverse distance method, identify the missing data at the point marked 'Z.'

Table 2.5   Station and rainfall data (inches)

Station	Annual Rainfall	April 2024 Rainfall
W	53.03	4.6
X	45.35	1.39
Y	59.73	5.83
Z	46.30	?

18. Compute the missing rainfall data for the station 400+00 for December 2022 using the record in Table 2.6. Assume the stations are approximately equidistant.

Table 2.6   Annual rainfall data (inches) for December 2022

Stations	Normal Annual Rainfall	Rainfall Data
100+00	16.02	9.29
200+00	15.43	6.06
300+00	18.91	10.3
400+00	14.45	?

19. Assume the normal annual rainfall at stations 100+00, 200+00, 300+00, and 400+00 in a basin are 80.97 cm, 67.59 cm, 76.28 cm, and 92.01 cm, respectively. In 2021, station 400+00 was inoperative, while stations 100+00, 200+00, and 300+00 recorded

annual rainfall of 91.11 cm, 72.23 cm, and 79.89 cm, respectively. Estimate the rainfall at station 400+00 for that year (see Table 2.7).

Table 2.7 Normal annual rainfall data (cm) for 2021

Stations	Normal Annual Rainfall	Rainfall Data
100+00	80.97	91.11
200+00	67.59	72.23
300+00	76.28	79.89
400+00	92.01	?

20. The execution of a riverbank protection project necessitates the extensive use of live vegetation and woody material, including pole planting through bioengineering methods. However, the proposed planting will not withstand the design storm until it is fully established. The designer is tasked with calculating the design storm and integrating temporary sediment control reinforcement matting into the design, ensuring a 90% probability of success over the next five years.

21. For station 100+00, Table 2.8 provides the recorded annual maximum rainfall over 24 hours. Compute the maximum 24-hour rainfall for return periods of 10, 25, and 50 years.

Table 2.8 Maximum 24-hour rainfall at station 100+00

Year	Rainfall (cm)	Year	Rainfall (cm)	Year	Rainfall (cm)
1995	14.08	2002	13.53	2009	9.32
1996	13.25	2003	12.36	2010	8.23
1997	6.09	2004	9.07	2011	6.51
1998	15.51	2005	9.29	2012	9.23
1999	15.93	2006	8.73	2013	11.11
2000	10.81	2007	8.98	2014	9.85
2001	7.56	2008	10.12	2015	9.23

22. What is the probability of a rainfall event of 10 cm or more occurring over 24 hours at station 100+00?

SELECTED SOURCES AND REFERENCES

Abu Hammad, A. H., A. A. Salameh, and R. Q. Fallah. 2022. "Precipitation Variability and Probabilities of Extreme Events in the Eastern Mediterranean Region (Latakia Governorate-Syria as a Case Study)." *Atmosphere* 13, no. 1: 131.

Afrin, S., M. M. Islam, and M. M. Rahman. 2015. "Development of IDF Curve for Dhaka City Based on Scaling Theory Under Future Precipitation Variability Due to Climate Change." *International Journal of Environmental Science and Development* 6, no. 5: 332–335.

Agustín, J. L. B., and R. D. López. 2010. "Efficient Design of Hybrid Renewable Energy Systems Using Evolutionary Algorithms." *Energy Conversion and Management* 50: 479–489.

Arndt, D. S., M. O. Baringer, and M. R. Johnson. 2010. "State of the Climate in 2009." *Bulletin of the American Meteorological Society*, 91(7): s1–s222.

Arnell, N. W. 2003. "Relative Effects of Multi-Decadal Climatic Variability and Changes in the Mean and Variability of Climate Due to Global Warming: Future Stream Flows in Britain." *Journal of Hydrology*, 270(3–4): 195–213.

Auffhammer, M., P. Baylis, and C. H. Hausman. 2017. "Climate Change Is Projected to Have Severe Impacts on the Frequency and Intensity of Peak Electricity Demand Across the United States." *Proceedings of the National Academy of Sciences*.

Bagheri, A., C. Zhao, and Y. Guo. 2017. "Data-Driven Chance-Constrained Stochastic Unit Commitment Under Wind Power Uncertainty." In *2017 IEEE Power & Energy Society General Meeting*, 1–5. https://doi.org/10.1109/PESGM.2017.8273948.

Baringo, A. and L. Baringo. 2017. "A Stochastic Adaptive Robust Optimization Approach for the Offering Strategy of a Virtual Power Plant." *IEEE Transactions on Power Systems* 32: 3492–3504.

Beyer, H.-G. and B. Sendhoff. 2007. "Robust Optimization—A Comprehensive Survey." *Computer Methods in Applied Mechanics and Engineering* 196: 3190–3218.

Blázquez, J. and A. S. Silvina. 2020. "Multiscale Precipitation Variability and Extremes over South America: Analysis of Future Changes from a Set of CORDEX Regional Climate Model Simulations." *Climate Dynamics* 55, no. 7: 2089–2106.

Celik, A. N. 2007. "Effect of Different Load Profiles on the Loss-of-Load Probability of Stand-Alone Photovoltaic Systems." *Renewable Energy* 32: 2096–2115.

Dowling, P. 2013. "The Impact of Climate Change on the European Energy System." *Energy Policy* 60: 406–417.

Evans, J. and S. Schreider. 2002. Hydrological Impacts of Climate Change on Inflows to Perth, Australia. *Climatic Change*, 55(3): 361–393.

Field, C. B., et al., eds. 2012. *Managing the Risks of Extreme Events and Disasters to Advance Climate Change Adaptation.* New York: Cambridge University Press.

Fisher, R. A. and L. H. C. Tippett. April, 1928. "Limiting Forms of the Frequency Distribution of the Largest or Smallest Member of a Sample." In *Mathematical Proceedings of the Cambridge Philosophical Society* (Vol. 24, no. 2, pp. 180–190). Cambridge University Press.

Fouskakis, D. and D. Draper. 2002. "Stochastic Optimization: A Review." *International Statistical Review* 70: 315–349.

Fréchet, M. 1927. Sur la loi de probabilité de lècart maximum. *Ann. Soc. Pol. Math. Crac.* 6: 93–117.

Gleason, K. L., et al. 2008. "A revised US climate extremes index." *Journal of Climate*, 21(10): 2124–2137.

Gnedenko, B. 1943. Sur la distribution limite du terme maximum d'une serie aleatoire. *Annals of Mathematics*, pp. 423–453.

Guntu, R. K. and A. Agarwal. 2020. "Investigation of Precipitation Variability and Extremes Using Information Theory." *Environmental Sciences Proceedings* 4, no. 1: 14.

Haigh, I. D., R. Nicholls, and N. Wells. 2010. "A Comparison of the Main Methods for Estimating Probabilities of Extreme Still Water Levels." *Coastal Engineering* 57, no. 9: 838–849.

Hall, I. J., R. R. Prairie, H. E. Anderson, and E. C. Boes. 1978. "Generation of a Typical Meteorological Year." In *Proceedings of the Conference: Analysis for Solar Heating and Cooling.* http://www.osti.gov/scitech/biblio/7013202.

Hasan, H. and Y. Wai Chung. 2010. "Extreme Value Modeling and Prediction of Extreme Rainfall: A Case Study of Penang." In *AIP Conference Proceedings* 1309, no. 1: 372–393.

Hawkins, E. and R. Sutton. 2009. "The Potential to Narrow Uncertainty in Regional Climate Predictions." *Bulletin of the American Meteorological Society* 90, no. 8: 1095–1108.

Hellström, C., D. Chen, C. Achberger, and J. Räisänen. 2001. "Comparison of Climate Change Scenarios for Sweden Based on Statistical and Dynamical Downscaling of Monthly Precipitation." *Climate Research* 19: 45–55.

Herrera, M., et al. 2017. "A Review of Current and Future Weather Data for Building Simulation." *Building Services Engineering Research and Technology* 38: 602–627.

Intergovernmental Panel on Climate Change (IPCC). 2007. World Meteorological Organization, CHANGE, O.C., 52: 1–43.

Intergovernmental Panel on Climate Change (IPCC). 2012. *Managing the Risks of Extreme Events and Disasters to Advance Climate Change Adaptation: Special Report of the Intergovernmental Panel on Climate Change.* Cambridge University Press. https://doi.org/10.1017/CBO9781139177245.

Intergovernmental Panel on Climate Change (IPCC). 2013. *Climate Change 2013: The Physical Science Basis.* Edited by T. F. Stocker, et al. Cambridge University Press.

Intergovernmental Panel on Climate Change (IPCC). 2014. *Climate Change 2014: Synthesis Report.* Edited by Core Writing Team, R. K. Pachauri and L. A. Meyer.

Jesan, T., C. Manonmani, S. Rajaram, P. M. Ravi, R. M. Tripathi, and S. Predeepkumar. 2016. "Statistical Analysis of Rainfall Events at Kalpakkam." In *3rd National Conference on Reliability and Safety Engineering (NCRS-2016).* SSN College, Chennai.

Ji, L., D. X. Niu, and G. H. Huang. 2014. "An Inexact Two-Stage Stochastic Robust Programming for Residential Micro-Grid Management-Based on Random Demand." *Energy* 67: 186–199.

Karl, T. R., et al., eds. 2008. *Weather and Climate Extremes in a Changing Climate. Regions of Focus: North America, Hawaii, the Caribbean, and the U.S. Pacific Islands.* Department of Commerce, National Climatic Data Center, Washington, D.C.

Karl, T. R., J. M. Melillo, and T. C. Peterson, eds. 2009. *Global Climate Change Impacts in the United States.* Cambridge University Press, New York.

Kay, A. L., N. S. Reynard, and R. G. Jones. 2006. "RCM rainfall for UK Flood Frequency Estimation. I. Method and Validation." *Journal of Hydrology*, 318: 151–162.

Kleinen, T. and G. Petschel-Held. 2007. "Integrated Assessment of Changes in Flooding Probabilities Due to Climate Change." *Climatic Change*, 81(3): 283–312.

Klemes, V. 1988. "The Improbable Probabilities of Extreme Floods and Droughts." In *Hydrology of Disasters: Proceedings of the World Meteorological Organization Technical Conference Held in Geneva*, November.

Kundzewicz, Z. W., U. Ulbrich and T. Brücher, et al. 2005. "Summer Floods in Central Europe—Climate Change Track?" *Natural Hazards* 36: 165–189. https://doi.org/10.1007/s11069-004-4547-6.

Kunkel, K. E., S. A. Changnon, and R. T. Shealy. 1993. "Temporal and Spatial Characteristics of Heavy-Precipitation Events in the Midwest." *Monthly Weather Review* 121, no. 3: 858–866.

LeBaron, B. and R. Samanta. 2005. "Extreme Value Theory and Fat Tails in Equity Markets." SSRN. https://ssrn.com/abstract=873656.

Lehner, B., P. Döll, J. Alcamo, T. Henrichs, and F. Kaspar. 2006. "Estimating the Impact of Global Change on Flood and Drought Risks in Europe: A Continental, Integrated Analysis." *Climatic Change*, 75: 273–299.

Lubchenco, J. and T. R. Karl. 2012. "Extreme Weather Events." *Physics Today*, 65, no. 3: 31.

Mavromatidis, G., K. Orehounig, and J. Carmeliet. 2018. "Comparison of Alternative Decision-Making Criteria in a Two-Stage Stochastic Program for the Design of Distributed Energy Systems Under Uncertainty." *Energy* 156: 709–724.

Medina, J. 2011. "Santa Ana Winds Buffet California." *New York Times*, December 2, A27. http://www.nytimes.com/2011/12/02/us/santa-ana-winds-buffet-california.html.

Minguez, R. and M. Menendez. 2012. "Statistical Models to Evaluate Long-Term Trends of Extreme Sea Level Events." In *AGU Fall Meeting Abstracts*, December, GC21G-06.

Mises, R.V. 1936. La distribution de la plus grande de n valeurs. *Rev. Math. Union Interbalcanique*, 1: 141–160.

Mote, P. W., E. A. Parson, A. F. Hamlet, W. S. Keeton, D. Lettenmaier, N. Mantua, E. L. Miles, D. W. Peterson, D. L. Peterson, R. Slaughter, and A. K. Snover. 2003. "Preparing for Climatic Change: The Water, Salmon, and Forests of the Pacific Northwest." *Climatic Change*, 61: 45–88.

Munich Re. 2012. "2011 Natural Catastrophe Year in Review." https://www.iii.org/sites/default/files/docs/pdf/MunichRe-010412.pdf.

Narayan, A. and K. Ponnambalam. 2017. "Risk-Averse Stochastic Programming Approach for Microgrid Planning Under Uncertainty." *Renewable Energy* 101: 399–408.

National Integrated Drought Information System. "U.S. Drought Portal." http://www.drought.gov.

Nguyen, T. H. and V. T. V. Nguyen. 2018, May. "A Novel Scale-Invariance Generalized Extreme Value Model Based on Probability Weighted Moments for Estimating Extreme Design Rainfalls in the Context of Climate Change. In World Environmental and Water Resources Congress 2018, pp. 251–261. Reston, VA: American Society of Civil Engineers.

Nik, V. M. 2012. "Hygrothermal Simulations of Buildings Concerning Uncertainties of the Future Climate." PhD diss., Chalmers University of Technology.

Nik, V. M. 2016. "Making Energy Simulation Easier for Future Climate—Synthesizing Typical and Extreme Weather Data Sets Out of Regional Climate Models (RCMs)." *Applied Energy*, 177: 204–226.

Nik, V. M. and A. Sasic Kalagasidis. 2013. "Impact Study of the Climate Change on the Energy Performance of the Building Stock in Stockholm Considering Four Climate Uncertainties." *Building and Environment* 60: 291–304.

Nkrumah, S. 2017. "Extreme Value Analysis of Temperature and Rainfall: Case Study of Some Selected Regions in Ghana." PhD diss., University of Ghana.

O'Brien, J. P., T. A. O'Brien, C. M. Patricola, and S. Y. S. Wang. 2019. "Metrics for Understanding Large-Scale Controls of Multivariate Temperature and Precipitation Variability." *Climate Dynamics* 53, no. 7: 3805–3823.

Palmer, T. N. and J. Räisänen 2002. Quantifying the Risk of Extreme Seasonal Precipitation Events in a Changing Climate, *Nature* 415: 512–514.

Panteli, M. and P. Mancarella. 2015. "Influence of Extreme Weather and Climate Change on the Resilience of Power Systems: Impacts and Possible Mitigation Strategies." *Electric Power Systems Research* 127: 259–270.

Patel, P. V., S. S. Patil, M. R. Kulkarni, A. G. Hegde, and R. P. Gurg. 2001. "Extreme Value Analysis of Meteorological Parameters Observed During the Period 1961–2000 at Tarapur." BARC–2001/E/025. Bhabha Atomic Research Centre.

Perera, A. T. D., R. A. Attalage, K. K. C. K. Perera, and V. P. C. Dassanayake. 2013. "Converting Existing Internal Combustion Generator (ICG) Systems into HESs in Standalone Applications." *Energy Conversion and Management* 74: 237–248.

Perera, A. T. D., V. M. Nik, D. Chen, et al. 2020. "Quantifying the Impacts of Climate Change and Extreme Climate Events on Energy Systems." *Nature Energy* 5: 150–159. https://doi.org/10.1038/s415 60-020-0558-0.

Perera, A. T. D., V. M. Nik, D. Mauree, and J.-L. Scartezzini. 2017a. "An Integrated Approach to Design Site Specific Distributed Electrical Hubs Combining Optimization, Multi-Criterion Assessment and Decision Making." *Energy* 134: 103–120.

Perera, A. T. D., V. M. Nik, D. Mauree, and J.-L. Scartezzini. 2017b. "Electrical Hubs: An Effective Way to Integrate Non-Dispatchable Renewable Energy Sources with Minimum Impact to the Grid." *Applied Energy* 190: 232–248.

Pryor, S. C., J. T. Schoof, and R. J. Barthelmie. 2006. "Winds of Change?: Projections of Near-Surface Winds Under Climate Change Scenarios." *Geophysical Research Letters* 33: L11702.

Sailor, D. J., M. Smith, and M. Hart. "Climate Change Implications for Wind Power Resources in the Northwest United States." *Renewable Energy*, 33(11), 2393–2406.

Samuelsson, P., et al. 2015. *The Surface Processes of the Rossby Centre Regional Atmospheric Climate Model (RCA4)*. Swedish Meteorological and Hydrological Institute. https://www.smhi.se/en/public ations/the-surfaceprocesses-of-the-rossby-centreregional-atmosp heric-climate-modelrca4-1.89801.

Santos, M. and M. Fragoso. 2013. "Precipitation Variability in Northern Portugal: Data Homogeneity Assessment and Trends in Extreme Precipitation Indices." *Atmospheric Research* 131: 34–45.

Sharifi, A. and Y. Yamagata. 2016. "Principles and Criteria for Assessing Urban Energy Resilience: A Literature Review." *Renewable and Sustainable Energy Reviews* 60: 1654–1677.

Smith, A. B. and R. W. Katz. 2013. U.S. Billion-Dollar Weather and Climate Disasters: Data Sources, Trends, Accuracy and Biases. *Natural Hazards,* 67(2): 387–410.

Solomon, S. D., et al., eds. 2007. *Climate Change 2007: The Physical Science Basis—Contribution of Working Group I to the Fourth Assessment Report of the Intergovernmental Panel on Climate Change.* Cambridge University Press, New York, p. 996.

Solomon, S. D. 2007. December. IPCC (2007): Climate change the physical science basis. In Agu fall meeting abstracts (Vol. 2007, pp. U43D-01).

Soroudi, A. and T. Amraee. 2013. "Decision Making Under Uncertainty in Energy Systems: State of the Art." *Renewable and Sustainable Energy Reviews* 28: 376–384.

Stern, P. C., B. K. Sovacool, and T. Dietz. 2016. "Towards a Science of Climate and Energy Choices." *Nature Climate Change* 6: 547–555.

Subramanya, K. 2006. *Engineering Hydrology.* Tata McGraw Hill Education Pvt. Ltd., New Delhi, pp. 1–387.

Tebaldi, C. and R. Knutti. 2007. "The Use of the Multi-Model Ensemble in Probabilistic Climate Projections." *Philosophical Transactions of the Royal Society A: Mathematical, Physical and Engineering Sciences*, 365(1857): 2053–2075.

United Nations, Department of Economic and Social Affairs. 2019. *World Urbanization Prospects: The 2018 Revision.*

U.S. Global Change Research Program. "National Climate Assessment." http://www.globalchange.gov/what-we-do/assessment.

Valdés-Pineda, R., J. B. Valdes, H. F. Diaz, and R. Pizarro-Tapia. 2016. "Analysis of Spatio-Temporal Changes in Annual and Seasonal Precipitation Variability in South America-Chile and Related Ocean–Atmosphere Circulation Patterns." *International Journal of Climatology* 36, no. 8: 2979–3001.

van der Wiel, K. and R. Bintanja. 2021. "Contribution of Climatic Changes in Mean and Variability to Monthly Temperature and Precipitation Extremes." *Communications Earth & Environment* 2, no. 1: 1–11.

Varakhedkar, V. K., B. Dube, and R. P. Gurg. 2002. "Extreme Value Analysis of Meteorological Parameters Observed at Narora during the Period 1989–2001" (No. BARC-2002/E/020). Bhabha Atomic Research Centre.

Wang, B., R.-Y. Ke, X.-C. Yuan, and Y.-M. Wei. 2014. "China's Regional Assessment of Renewable Energy Vulnerability to Climate Change." *Renewable and Sustainable Energy Reviews* 40: 185–195.

Wilby, R. L. and T. M. L. Wigley. 2002. Future Changes in the Distribution of Daily Precipitation Totals Across North America. *Geophysical Research Letters*, 29(7): 39–1.

Wilby, R. L., C. W. Dawson, and E. M. Barrow. 2002. SDSM—a decision support tool for the assessment of regional climate change impacts. *Environmental Modelling & Software*, 17(2): 145–157.

World Economic Forum. 2016. *The Global Risks Report 2016.* https://www.weforum.org/reports/the-global-risks-report-2016/.

Zanocco, C., H. Boudet, R. Nilson, H. Satein, H. Whitley, and J. Flora. 2018. "Place, Proximity, and Perceived Harm: Extreme Weather Events and Views about Climate Change." *Climatic Change* 149, no. 3: 349–365.

Zawiah, W. Z. W., A. A. Jemain, K. Ibrahim, J. Suhaila, and M. D. Sayang. 2009. "A Comparative Study of Extreme Rainfall in Peninsular Malaysia: With Reference to Partial Duration and Annual Extreme Series." *Sains Malaysiana* 38, no. 5: 751–760.

Zhao, C. and Y. Guan. 2013. "Unified Stochastic and Robust Unit Commitment." *IEEE Transactions on Power Systems* 28: 3353–3361.

3

FUNDAMENTALS OF
URBAN DRAINAGE
CONVENIENCE SYSTEMS

3.1 INTRODUCTION

Urban drainage infrastructures are designed to effectively collect, transport, and control stormwater runoff resulting from precipitation or snowmelt. These systems comprise a variety of structures and components, including catch basins, manholes, levees, spillways, detention and retention structures, service lateral lines, pipes, ditches, culverts, swales, subsurface storage, and roadside curbs and gutters. Each element is designed to fulfill a specific function, working in unison to move, store, and treat stormwater runoff. This collaboration helps mitigate the potential impact of stormwater runoff and prevents flooding.

Designing storm drainage projects requires careful consideration of engineering principles and the potential impacts of extreme weather events. Successful implementation of these projects necessitates the use of advanced hydrologic and hydraulic modeling techniques. Thus, a comprehensive understanding of two fundamental engineering disciplines is essential for the successful planning and construction of stormwater drainage systems:

1. *Hydrology*—which involves the scientific measurement and analysis of precipitation and snowmelt that contributes to surface water runoff (it also probes the various impacting elements in-depth).

2. *Hydraulics in water resource engineering*—which requires a thorough examination and assessment of various factors associated with the flow of stormwater runoff in diverse conveyance systems, including channels, pipes, streams, ponds, and rivers.

3.1.1 The Evolution of Urban Drainage Systems

Various elements have influenced urban drainage systems over the course of city development. Urban drainage systems have undergone significant advancements from ancient civilizations such as the Indus and the Minoan to the modern design methods and practices used today. Urban drainage systems are essential for effectively managing surface stormwater runoff. This section will explore the history and evolution of drainage systems as well as examine potential future trends and innovations. An extensive range of urban drainage systems that are not only fascinating but also peculiar can be found, and historical sources offer insights into the processes that are involved in these systems. For example, Lewis Mumford (1937), an American historian, sociologist, philosopher of technology, and literary critic, gave a brief overview of the state of urban infrastructure in ancient times. It is possible to trace the origins of design standards, practical formulas, scientific concepts, mathematical methods, and modeling in hydrology and hydraulics all the way back to ancient engineering. The development of these components has occurred over the course of the history of modern engineering, and they have gained widespread acceptance and application. In spite of the fact that contemporary engineers have a major technological edge over their predecessors, there have been several instances throughout history in which ancient engineers have accomplished remarkable feats equal to those of present-day engineers.

A wide range of ancient civilizations used materials like stone, brick, and wood to create absolutely stunning buildings. Roadways, water supply and distribution systems, wastewater collection systems, and stormwater drainage systems were some of the cutting-edge infrastructure systems that were installed in these communities. The collection of wastewater and the drainage of stormwater are two examples of the types of infrastructure systems that were frequently linked together. *Wastewater*, which is simply water that has been used for the support of life, industrial operations, or life improvement, must be collected and disposed of properly.

In the context of water management, the term *stormwater* refers to the runoff that is produced as a result of precipitation. Both wastewater and stormwater must be taken into consideration throughout the planning process for urban drainage systems. This is a vital step in the process. Throughout the course of history, the two waters have either been mixed into a single conduit (also known as combined sewers) or have been maintained separately during the collection and disposal procedures (also known as separate sewers).

3.1.2 Exploring Ancient Urban Drainage Systems

Based on the historical accounts of ancient civilizations like the Indus and the Minoan, it can be inferred that urban drainage systems were developed with a great deal of attention to detail. These systems were designed to collect rainwater, prevent nuisance floods, and transport rubbish. It is quite probable that the systems that eventually accomplished their objectives did so after going through a process of trial and error, which involved making modifications to the system. With relation to urban drainage, there were not a great deal of numerical standards available, and engineering calculations were not applied during the design process. Despite the fact that there was no optimization process and building methods were dependent on trial and error, there are a vast number of older urban drainage systems that can be considered to be highly successful. The ancient sewer networks were an uneconomic combination of sophisticated technological instruments and unsophisticated social design. Lewis Mumford, who was also a philosopher of technology, provided a concise summary of the condition of urban infrastructure in ancient times.

The Indus civilization flourished in the valley of the Indus River during the beginning of the third millennium BCE. Several scholars, including Webster (1962) and Kirby et al. (1956), have brought to light the fact that the Indus civilization was responsible for the creation of complex urban drainage systems for a number of its most important cities. The urban drainage systems that the Indus civilization utilized were made abundantly obvious by the crumbling remnants of two particular cities. Harappa and Mohenjo-Daro, two cities from the Indus Valley civilization, were located about 350 miles apart. These urban centers had well-planned construction and integrated an advanced municipal drainage system into their layout. The majority of the homes in these settlements were connected by an open channel constructed in the middle of the road, which was made of either regular burned bricks laid in clay mortar or excavated earth (for additional details, see Figure 3.1).

Although houses were connected to the drainage channels, wastewater was not permitted to flow directly to the street sewers. Instead, they routed the wastewater through tapered clay or terra cotta sewer pipes into a small sump to settle and accumulate solids before overflowing into street drainage channels. The drainage channels were covered by bricks and cut stones, which could be removed during maintenance and cleaning activities. Overall, the Indus civilization viewed urban drainage systems as serving two purposes: waste and stormwater conveyance.

Figure 3.1 Early urban drainage system (Hodge 1992 and Kirby et al. 1956)

The inclusion of a cunnette, which is a small channel dug into the bottom of a larger channel or conduit, served to concentrate the flow during low-water stages. This was initially described by Webster in 1962 and was yet another fascinating component of the channel that was in existence at the time. The cunnette was most likely created to transfer the smaller flows that were associated with the daily discharges of wastewater. When there was rain, the remaining portion of the waterway would be used for transportation purposes. During the time of the Indus civilization, it was believed that urban drainage networks had the twin goal of transporting waste and providing drainage for stormwater. In a general sense, this concept was known to many. In addition to other ancient cultures, the Persians also built drainage systems for urban areas. According to Niemczynowicz (1997), the ancient Persians held the view that urban runoff was sacred and imposed restrictions to prevent it from becoming polluted.

In ancient Persia, polluting the water supply was deemed inappropriate and offensive behavior. In addition, cisterns were used to collect rainwater and runoff from urban areas with the aim of making it drinkable. The urban runoff was injected into the underlying aquifer through the use of deep wells. From the Persian point of view, urban runoff was unquestionably seen as an immensely important natural resource. According to Niemczynowicz (1997), the progression of time brought about a change in the attitudes and behaviors of the Persian people, which ultimately led to the problems of water pollution and the eventual collapse of the civilization.

Both Assyria and Babylonia, which were realms of the Mesopotamian Empire, made significant contributions to the development of civilization during the second millennium BCE. The ruins of Mesopotamian cities have been found to feature storm drainage and sanitary sewer systems that were exceptionally well designed. As an illustration, the ancient towns of Ur and Babylon, which are situated in what is now Iraq, were equipped with efficient drainage systems to control stormwater. When it came to the disposal of household garbage, the systems included vaulted sewers and drains, as well as gutters and drains that were specifically built to deal with surface runoff. Brick that had been baked and then sealed with asphalt was selected as the material of choice.

In addition, rainwater was collected for use in irrigation systems that were installed within a residence. The need to maintain a clean environment was one of many factors that influenced the design of urban drainage systems in Babylonia. Like other ancient civilizations, the Babylonians considered uncleanliness to be a socially unacceptable and forbidden habit, not because it signified a lack of cleanliness in the physical sense, but rather because it symbolized a moral illness. According to Mesopotamians, urban runoff was not only a source of garbage but also a source of floods, and they felt it to be an annoyance. Despite that, it was also regarded as a necessary natural resource.

The Minoan civilization existed on the island of Crete between the years 2800 and 1100 BCE. The relics of this civilization, which was situated on the Aegean Sea, were found to have revealed intricate networks of stone drains that had been constructed with considerable care, as stated by Gray (1940) (see Figure 3.2).

The construction of these drains was done in layers. Prior to their discontinuation, these drains were used to transport a variety of different types of drainage, including sanitary sewage, roof runoff, and even ordinary surface drainage. In order to dispose of the sewage, the

Figure 3.2   Minoan storm drain located in Knossos, dating back to 1500 BCE

drains discharged their contents into a major sewer, which was located a considerable distance away from the origin of the waste. Eventually, the contents of the drains were discharged into the sewer. The system was pushed out to the required degree as a result of the widespread and intense rainfall that occurred on the island of Crete many years ago. In the course of the excavations that have been carried out in the royal city of Knossos, it has been discovered that a system was developed that was comprised of two conduits.

One of the conduits was used to collect rainwater, while the other was used to collect sewage. It is possible that the Minoans considered urban runoff to be a cause of flooding, a means of conveying garbage, and an essential natural resource. It is crucial to highlight that this theory is supported by the presence of a unique and effective urban drainage system that was specifically created to incorporate stormwater collection equipment.

The Etruscan civilization was responsible for the development of some of the oldest organized settlements in central Italy. These settlements were likely constructed around 600 BCE. A drainage system that was properly constructed and took advantage of the natural slope of the land was installed in Marzacotto, which was one of the more major Etruscan cities. This system was designed to keep the environment within the city dry and clean. Furthermore, an additional kind of protection for pedestrians against the discharge of stormwater was provided by cemented pavements and stepping stones that were placed on streets (Strong 1968). As with other ancient civilizations, the Etruscans saw urban runoff as a cause of floods, a vehicle for the transportation of waste, and a resource that was essential to urban development.

3.1.3 Urban Drainage Systems in the Roman Empire

Significant advancements were achieved in urban drainage during and after the Roman Empire. One example that stands out is the standardization of drainage methods for roads and the construction of large underground pipes linked to each other to make a complex network of drains. The next section will discuss urban drainage from the Roman point of view, which was mostly about keeping plains from flooding and, if necessary, draining them. However, collecting rainwater for use in homes and public places was also seen as very important.

Across different eras, from ancient civilizations to the 19th century, the Romans were distinguished by their exceptional road network in western Asia and Europe. Their roads were known for their meticulous design and well-maintained surfaces (see Figure 3.3). Between 800 and 350 BCE, the Etruscans held power in Italy. In that era, a significant number of roadways were constructed. Nevertheless, the roads did not possess the same intricate design as Roman roads, and the drainage systems were not as meticulously planned.

Following their invasion of the region, the Romans improved and extended many of the Etruscan pipelines and roadways. After thorough analysis, it has been determined that curbs and gutters were installed

Figure 3.3   Roadways built during the Etruscan rule in Italy between 800 and 350 BCE

on specific streets to redirect surface runoff into open drainage ditches lined with pebbles. To divert the surface runoff from the streets into the drainage channels, a substantial part of the roadbeds needed to be graded. Furthermore, rainwater collection was a key element in the Romans' drainage system, complementing the urban drainage component integrated into road construction. Rainfall in urban areas is typically collected and stored for local use. Rainwater that landed on rooftops was often stored in a cistern located indoors.

The Cloaca Maxima is the largest of the Roman sewers, in addition to being the oldest and most prominent structure in the cloaca (sewer) network (see Figure 3.4). It was responsible for draining the lowest areas of Rome, which were located around the Forum, into the Tiber River (Gest 1963). Both the city and the historic structures of the Roman Forum and beyond are buried beneath it. In the early days of the Roman Forum, the valley in which it was situated was a wetland, which made it difficult to build anything in the area. Rome continued to use the *Great Drain* as its principal sewer system, and it is still used to carry out some of the functions for which it was originally designed.

Figure 3.4   Cloaca Maxima—the *Great Drain*

3.1.4 Post-Roman Empire Drainage Systems

As individuals migrated away from urban areas after the decline of the Roman Empire, the populations of cities in Europe and parts of Asia started to decrease significantly. Due to a decrease in population in urban areas, various municipal services, such as sewage systems and running water, were discontinued or degraded due to a lack of attention. Urban residents displayed a lack of concern and disinterest during the Dark Ages, which also played a role in the deterioration of urban drainage systems at that time (Bishop 1968). The Dark Ages were a period of time in Europe during which only a limited number of technological improvements were devised, and even those were rarely put into practice. As a direct result of this, certain components of urban infrastructure, such as urban drainage systems, were not being brought up-to-date. When it came to urban drainage, the dominant perception held by the general population during this time period was that it was a service that was not required.

The urban stormwater runoff and the industrial wastewater from the processes employed by tanners and dyers were the primary sources of rubbish that entered the local streams and rivers where it was eventually deposited. The use of human waste as fertilizer in home gardens was common. Pigs were frequently given a wide variety of waste products and garbage from households, which was typically stacked up close to the city for storage. Following the decline of the Roman Empire, the sewers in Europe were simply open ditches designed to gather stormwater. However, despite being primarily intended for stormwater drainage in urban areas, the open ditches also accumulated garbage and waste, which obstructed the flow of stormwater and caused flooding. Eventually, the open passages were covered in order to combat the sanitation problems that were developing and becoming a nuisance. This construction is believed to be the earliest covered sewer that was constructed in a city (Reid 1991).

The cities of Paris and London saw a gradual rise in population toward the end of the Dark Ages, which resulted in the emergence of a problem regarding the disposal of human feces. A 1530 decree required the construction of cesspools in every new residence in Paris (Reid 1991). This edict was addressed to private property owners. Prior to the issuance of that order, the majority of Paris's rubbish disposal was not controlled. From a general standpoint, every town and neighborhood had a self-centered attitude with regard to urban drainage and the services that were provided by the municipality. There was a willingness among the public to pay for sewers that would drain only their neighborhood into the next community.

Throughout the Middle Ages, covered sewers in Europe were plagued by a multitude of maintenance problems. An investigation into the condition of Paris's covered sewers was carried out in 1636, and the results showed that all 24 of them were clogged and in a state of severe degradation (Krupa 1991). Following those findings, there was very little progress made because the ruling elite and the nobles showed little care for the well-being of the ordinary people. Property owners were required to pay for the cleaning of the sewers that were installed beneath their buildings in accordance with a regulation that was passed in 1721. Since property owners believed they could dispose of all of their trash in the sewage system as long as they paid for it, this action made the problem of urban drainage much worse. In 1736 and 1755, a number of further rules were enacted in an effort to discourage the act of depositing waste into covered sewers without authorization. However, these efforts were not well embraced.

The year 1427 was significant for England since it was the year that one of the earliest public acts was passed to address sewer difficulties (Sidwick 1977). To regulate surface water sewers and channels, the Commissions of Sewers Act was passed. The Act was amended in 1531, and it continued to be in force until 1848 when the Public Health Act was finally written and signed into law. Regrettably, the implementation of the Act proved to be difficult, even though the Act of 1427 brought attention to the growing concern of the ruling class regarding sanitation.

According to Gayman (1997), King Henry VIII issued an order that every household was required to clean the sewer that flowed by their residence. This decree was similar to an ordinance that was passed in Paris. By 1622, an agreement had been made to employ fines for noncompliance as a means of funding the activities of the organization. This was because there was no money set aside to compensate its members for their involvement in the commissions of sewers. Although there was a minor improvement in the ruling class's and society's outlook on urban drainage difficulties at this time, the systems mostly continued to be ignored.

Over the course of our exploration of the Americas, we have observed a shift in emphasis toward the transformation of urban drainage into public works systems that require careful planning, construction, and maintenance. The Incas, who lived in South America, possessed a deep understanding of the relevance of urban drainage technologies. One particularly noteworthy example is the drainage system that was installed in Machu Picchu, which was the royal estate of Pachauri, the Incan king. A comprehensive investigation of the remains of the estate was carried out by Wright and Valencia Zegarra in 2000. The investigation focused on the engineering components of the drainage system. This drainage system at Machu Picchu was deliberately built utilizing the enormous amounts of information that the Incan civilization had regarding the development of drainage infrastructure from earlier settlements through trial and error. The fact that the Incas placed a high priority on urban drainage can be seen from the painstaking planning and construction of a sophisticated drainage system within the city boundary.

3.1.5 Large-Scale Drainage Systems in North America

According to the American Public Works Association (APWA) (1976), the first large urban drainage systems in North America were built in New England settlements throughout the eras of colonial control. Draining the roads was an essential step that needed to be taken in order to make sure that horse-drawn carriages could use them. At the beginning of the 18th century, various cities—including Boston, Philadelphia, and New York—developed stone roadways that included both surface and subsurface drainage systems. An exceptionally high level of forethought and attention to detail was displayed in the design of the highways in Boston, which had gutters running along the sides and a crown in the middle of the roadways. Stone, brick, and wood were some of the materials that were used in the construction of the first sewers in New England during the colonial era (APWA 1976). These sewers were constructed in the commercial areas of the major cities in New England. Private sewers made of wood were built in the latter part of the 17th century to drain cellars. APWA (1976) states that in 1704, Francis Thrasher, a resident of Boston, was granted permission to construct a sewer that would be of use to the general public; it was among the first common sewers that were constructed lawfully in New England. As a consequence, the local selectman issued a mandate requiring those who were to be connected to the sewer system to offer financial support for the project.

In the 18th century, Boston launched the construction of the first underground sewer system. This was a project that had popular support from the people who lived in the city. As soon as it was realized that the project had been successful, a number of other sewage projects were undertaken. The APWA reports that between 1708 and 1736, Boston issued a total of 654 sewer-building licenses for projects that required meticulous roadway repair upon completion. One of the reasons that Boston is considered to be one of the cleanest and driest cities in the world is because of the massive sewer infrastructure that has been established throughout the city.

Philadelphia also conducted a significant amount of construction work on its sewer systems (see Figure 3.5). A substantial number of the sewers that were built in Philadelphia throughout the 17th century were made of wood, which was the material of choice at that time. In the 1700s, Pennsylvania became the first state to pass a law requiring the use of brick or stone for underground sewer construction. However, within a short period of time, the newly constructed sewers became clogged with solid material, and as time went on, they gradually began to deteriorate due to the difficulty of proper municipal maintenance.

Figure 3.6 shows the construction of the sewer system in Norwood Park, Chicago, around 1916. This perspective reveals the finished invert, arch forms, bracing and sheeting, layers of brick, and the initial

Figure 3.5 Mill Creek Sewer construction (47th and Haverford; Philadelphia Water Department 1883)

Figure 3.6 Construction of Chicago Norwood Park sewer, circa 1916

stage of trench filling—from page 177 of the book *Chicago's Norwood Park Sewer: Methods of Construction*, which was published in November 1916 in the volume of Municipal Engineering, volume LI, number 5. In general, the theory of urban drainage in the cities of America was remarkably similar to the procedures that were used in Europe.

3.1.6 Modern Development of Large-Scale Drainage Systems

In the first part of the nineteenth century, the sewage system of Paris became known for being a haven for criminals and other undesirables (see Figure 3.7). This reputation persisted until quite recently. In his article from 1991, Reid made the argument that the relative relationship between the underground sewer system and the above-ground city was emblematic of the class fights that were occurring during that time period. In the 19th century, the city of Paris employed an engineer named Eugene Belgrand. He made substantial enhancements to

Figure 3.7 Paris sewer system

the sewer system by installing more than 1,000 kilometers (621 miles) of new sewers, expanding the stormwater drains and roads, establishing a waste treatment plant, and constructing aqueducts to supply Paris with drinking water from the nearby area.

Significant advancements were achieved in improving the waste management system in Paris during this time period. Expanding the sewers led to a reduction in disease and pollution, ultimately resulting in decreased mortality rates. General key urban drainage developments that occurred over the nineteenth and twentieth centuries were outlined in a summary that was published by Burian et al. (1999). Nine categories were used to classify the developments:

1. Improvements in pipe materials, construction methods, and maintenance practices
2. Consideration of utilizing a water-carriage system for the removal of waste
3. A comprehensive approach to the design of sewer systems
4. Combined sewer system (CSS) versus Municipal Separate Storm Sewer System (MS4)
5. The identification of waterborne diseases
6. Innovative wastewater treatment strategies
7. Modern urban hydrology and hydraulics
8. Advancements in modeling technology
9. Growing awareness and prominence of the environment

Given the breadth of this chapter, it would be impossible to provide a complete overview of the historical developments that have occurred within each of these categories. Instead, we can provide a summary of the significant shifts in urban drainage perspectives that occurred over each period of time.

At the beginning of the nineteenth century, the process of waste disposal consisted of discharging hygienic waste into cesspools and privy vaults. The beginning of the 1900s saw the end of this long-standing custom. In most cases, the primary purpose of sewers was to facilitate the drainage of stormwater. In the past, garbage would accumulate in private vaults and cesspools, and then workers would come by regularly to collect it and transport it to a specified disposal place, such as a farm or a dump located outside of the city. As a result of the introduction of water-carrying sanitary waste collection systems in urban areas during the nineteenth century, urban drainage underwent a considerable transformation. In addition, new sewers were constructed in order to control stormwater and sanitary wastewater, and this also made it possible for sanitary connections to be made to the sewers.

Throughout the nineteenth century, there was a discernible shift in the way that the general public approached the topic of urban drainage. In the process of the development of public works, it went from being ignored to being acknowledged as an essential component with significant importance. There was a change in the way that the general public perceived the financing of the construction and maintenance of new sewer systems. However, the scientific evidence that emerged in the latter half of the century correlating sanitary wastes with disease transmission was likely the most important influence in transforming the public's understanding of the issue.

The leaders of large cities in Europe and the United States led huge campaigns to develop vast sewer networks and these initiatives gained a significant amount of support from the public. In 1843, the city of Hamburg, Germany, built one of the earliest complete sewerage systems for a major metropolis (Metcalf and Eddy 1928). Following the destruction of a significant portion of the city by fire in 1842, William Lindley was given the responsibility of planning and designing the new sewer structure. The comprehensive design of sewer systems in various cities across Europe and the United States was directly impacted by the sewer system that was in place in Hamburg. The huge London Main Drainage sewage construction project of 1859–1865 led by Joseph Bazalgette is depicted in Figure 3.8, which allows for a better understanding of the magnitude of the undertaking.

Furthermore, in the nineteenth century, there was a notable change in the perception of urban drainage from a design perspective. Many sewers built prior to the nineteenth century were not planned or designed by engineers. However, the pursuit of improvement yielded significant advancements; a systematic approach of trial and error led to

Figure 3.8   London's 1860's drainage system

the development of more efficient systems in certain cases. In the nineteenth century, there was a shift in the approach to urban drainage design. Technical experts began to play a significant role, and engineers started extensively using numerical calculations for planning and designing sewer networks. The design of the Hamburg sewer system, for example, incorporated precise engineering calculations (Metcalf and Eddy 1928). Other engineers developed their own design methods. Joseph Bazalgette, for example, determined the target threshold for combined sewer overflows by calculating the amount of rainfall runoff that would occur during a specific frequency event. Based on this information, he calculated the extra volume of pipe needed to ensure the system operates effectively. Charles Buerger (1915) provided a comprehensive overview of the sewer-sizing calculations that were prevalent in the early nineteenth century, along with a selection of widely used empirical equations.

In the latter half of the 19th century, Mulvaney (1851) in Ireland, Kuichling (1889) in the United States, and Lloyd-Davies (1906) in Great Britain all wrote about the calculation of runoff and the sizing of sewer pipes. They introduced the concept of time of concentration, which eventually led to the development of the rational method, which greatly improved the design of urban drainage and sewer systems. However, as water pollution and public health issues related to uncontrolled sewer discharges into waterways increased, the focus of urban drainage shifted to include and emphasize sewage treatment as a feasible solution.

In Europe and the United States, urban drainage treatment was largely inadequate in the late 1800s despite scientific studies demonstrating a direct link between sewage-contaminated waterways and the spread of disease. There were a mere 27 communities in the United States that possessed effluent treatment plants by 1892 (Tarr 1979). Of them, 21 were associated with land application, and six were associated with chemical precipitation. Many states and municipalities in the United States still prefer land-application wastewater treatment and disposal methods for treated wastewater in order to avoid the discharge of untreated wastewater into downstream waterbodies. Furthermore, land application is a beneficial re-use of these wastes and typically costs 30 to 50 percent less to operate than high-energy-demanding treatment options. The other option for wastewater treatment is chemical precipitation, which is the transformation of dissolved materials into solid particles. It eliminates contaminants from both municipal and industrial wastewaters. Despite these alternatives, the discussion on wastewater treatment was centered around the cost-effectiveness of treating the wastewater prior to disposal or treating the water source to make it suitable for drinking. At that time, most sanitation experts believed it was more equitable for a city located upstream to discharge its sewage into a stream and have the downstream city take responsibility for filtering the water for residential purposes than requiring the upstream city to construct wastewater treatment facilities.

3.1.7 Current Urban Drainage Practices

Throughout the twentieth century, urban drainage became a critical element of public works development. Engineers were consistently enhancing design concepts and methodologies. Regulatory measures were implemented in various locations around the world to address urban drainage issues. The use of computer modeling tools broadened the approaches employed in designing and analyzing sewer systems. Urban drainage evolved from a focus on public health and nuisance flooding in the early 1900s to now include ecosystem protection and urban sustainability. This shift in perspective has occurred as a result of regulations, monitoring, computer modeling, and environmental concerns.

3.1.8 Urban Drainage Systems—Emerging Trends and Innovations

Continuous research and evaluation help integrate cutting-edge urban drainage systems and enhance the development of resilient, livable, sustainable, and environmentally friendly cities. However, the challenges and opportunities in these sectors are constantly evolving due to climate change, population growth, urbanization, and technological innovation. Modern development concepts, such as low-impact development, best management practices (BMPs), and green infrastructure are shaping future development practices to reduce negative impacts on stormwater drainage. In certain scenarios, alternative on-site stormwater management solutions are being promoted as more environmentally friendly than centralized stormwater management. Communities are exploring new methods to gather, store, and make use of rainwater within the watershed instead of constructing extensive drainage systems. Several municipalities are in the process of creating stormwater quality management plans that cover the entire watershed to tackle flooding and water-quality issues. Metropolitan drainage has seen substantial expansion in recent years, evolving to address a wide range of factors including social, economic, political, environmental, and regulatory considerations.

3.1.9 Recognizing the Importance of Stormwater Management

When it comes to regulating the effects of urbanization on flow patterns, the conventional approach has been concentrated on reducing

the effects of increasing peak flows that are caused by uncommon occurrences (for example, a 100-year average recurrence interval event).
The reduction of dangers to human health as well as harm to property and infrastructure has been the overarching goal of this strategy.
The control of flows resulting from frequent events of relatively minor
magnitude is an area that has to be prioritized from the point of view
of environmental management. This is primarily due to the fact that in
urban areas, the most substantial changes to the hydrological regime
occur in the frequency and amplitude of the smaller runoff events that
occur more frequently.

The majority of the time, runoff source controls consist of things
like infiltration methods or stormwater reuse procedures. A wide variety of infiltration methods are available, ranging from those that are
deployed on individual housing blocks to those that are built into the
sewer system. These include the following methods that can help manage roof runoff effectively:

- Infiltration trenches, dry wells, and soak ways for infiltration
- Directing roof runoff to ponding areas in backyards for infil-
  tration
- Grassed swales
- Perforated pipes
- Porous pavements
- Infiltration trenches and basins within the drainage system

When applied in regions with soils that have a reasonably high permeability, these systems are excellent for infiltrating runoff that contains
modest quantities of sediment. It is possible to utilize grassed swales
in place of curbs and gutters since they have the ability to slow down
the flow of water around them. The failure rate of porous pavements is
substantially higher than average.

In order for infiltration techniques to be effective over the course
of a long period of time, they need to be constructed in a manner
that is efficient, applied in areas that have a low volume of traffic, and
utilized to infiltrate stormwater to the extent that there is a negligible amount of silt. Managers should take into consideration whether

or not pretreatment of the runoff is necessary in order to guarantee that these techniques will continue to perform effectively over the long term when picking the most appropriate infiltration technique. An additional factor that should be taken into consideration is the possibility of groundwater pollution or the escalation of salinity issues in metropolitan areas. To further improve the management of septic tank systems, sullage discharges, and sewerage system overflows, managers should also take into consideration the actions that can be implemented to improve management.

The reuse of stormwater can be an efficient method for controlling runoff, and it also has the added benefit of supplying an extra water resource. There are three different levels at which reuse measures can be implemented: the individual block level, the catchment level, and the subcatchment level. Rainwater tanks and other similar devices for collecting roof runoff, wet basins, and newly formed wetlands are examples of these types of structures. The water that is collected can be utilized for a variety of projects that do not require potable water, such as industrial activities, irrigation, watering gardens, and flushing toilets. Therefore, these devices should not be utilized as a replacement for any on-site detention requirements that are necessary for flood mitigation. This is because the volume of water that can be stored in these devices will vary. The storage, on the other hand, can be constructed to accommodate a variety of purposes. Both flow attenuation and the reuse of stormwater could be accomplished through the provision of storage.

3.1.10 Principles of Urban Storm Management

There are some complicated interactions that take place between the hydrological properties of a watercourse; these include streamflow, water quality, channel form, aquatic habitat, and riparian vegetation. In addition, these interactions will have an effect on human health, recreational activities, and aesthetic concerns. It is vital to incorporate a holistic strategy in order to ensure that these principles continue to be upheld over the long term. In a similar vein, it is highly improbable that the management of a single component of a stormwater system will accommodate all of the pertinent factors. When managing storm-water, it is important to take into account the hydrological, geomor-phological, biological, soil, land use, and cultural aspects of a watershed and the watercourse network that it contains. There is a possibility that well-intended management approaches could have a bigger impact on the ecosystem than unrestrained stormwater runoff if the interactions that were previously mentioned are not taken into consideration. The following is a list of general and comprehensive guidelines that can be adhered to in order to effectively manage the environment caused by stormwater:

1. *Hydrological*: The process of reducing the effects of urbaniza-tion on the hydrological features of a watershed, such as low flows and wet weather, is referred to as hydrological. If prede-velopment land use leads to an improper streamflow regime (for example, runoff from an agricultural catchment that causes erosion), this should be addressed whenever possible.
2. *Water quality*: To improve water quality, it is necessary to reduce the amount of pollution that is introduced into the stormwater system and to eliminate an adequate quantity of any pollution that is left behind through the use of stormwater management measures.
3. *Ecosystem diversity*: Both vegetation and aquatic habitat are important aspects of the stormwater system. Vegetation aims to maximize the value of native riparian, floodplain, and fore-shore vegetation, while aquatic habitat aims to maximize the value of physical habitats to aquatic species.

It is possible for the relative importance of these principles to dif-fer—both within and between catchments—despite the fact that they are highly connected. In order to reach a balanced conclusion that maximizes the overall environmental, social, and economic benefits, it may be necessary to make concessions between these principles at any particular location. On the other hand, given that runoff rates and volumes frequently, either directly or indirectly, produce a wide vari-ety of potential social, economic, and environmental repercussions in metropolitan settings, it is absolutely necessary to manage these particular runoff characteristics. The following examples describe recommended stormwater management approaches that can be discovered through the identification of a stormwater management hierarchy:

1. *Retain and restore (or rehabilitate) valuable ecosystems*: This includes maintaining or repairing (if degraded) existing significant features of the stormwater system, such as natural channels, wetlands, and riparian vegetation—in effect, rehabilitating such ecosystems.

2. *Nonstructural and structural source control*: Refers to nonstructural measures, which are approaches used to limit changes in the nature and quantity of stormwater at the source and structural measures, which are management strategies that are created and placed at or near the source.

3. *Online and offline configuration systems*: Various stormwater management techniques are implemented within stormwater systems to control the volume and quality of stormwater before it is released into receiving waters. Stormwater BMPs can be set up in either an online or offline peak flow configuration:

   - An *online configuration* is when a BMP is set up on the main line so that excess runoff flow rates can be directed through BMPs to ensure a consistent reduction in pollutants and total peak flow discharges.
   - An *offline configuration* is when peak flow rates are diverted around the BMPs and with only water quality flow passing through the BMPs using a diversion structure such as a weir installed either internally or externally. Typically, the vast majority of stormwater treatment BMPs are set up in an online configuration.

3.1.11 Managing Stormwater Runoff Effectively

Engineers and water authorities face a big issue when it comes to dealing with stormwater runoff in metropolitan areas. Historically, precipitation would gradually infiltrate into the earth, and it would eventually find its way into bodies of water such as rivers, lakes, and seas. As a result of human activity, woodlands and grasslands have been replaced by a multitude of impermeable surfaces, which has resulted in the rapid flow of runoff into neighboring waterways. The huge amounts of water and contaminants that are associated with stormwater runoff represent a substantial risk to natural ecosystem streams. The management of stormwater is typically a large-scale task that is commonly divided among several levels of government (state, local, and municipal). Although the majority of attention is often directed toward huge projects and storm drainage systems, it is important to note that even relatively small places can contribute significant volumes of rainwater during rainstorms. It is possible to achieve significantly higher levels of infiltration throughout the watershed region by implementing improvements at the residential lot level. Each homeowner has the ability to greatly lessen the amount of stormwater that is discharged from their property, which will result in an improvement in the quality of surface water and will also contribute to the replenishment of groundwater reserves.

Within the context of an integrated system, a number of BMPs will be utilized by an efficient stormwater management plan in order to lower the amount of pollution that is introduced into our waterways, thereby lowering the likelihood of floods. However, when evaluating BMPs for usage at a site, it is essential to do an analysis to determine which primary issues should be addressed prior to planning and establishing the BMPs due to the fact that these elements have the potential to influence the size of the site, the services it provides, and how it operates. BMPs differ in terms of their benefits, limitations, and qualities. These features decide whether or not they are appropriate for use in a specific context. Consider the scenario where we desire to manage stormwater flow with the following properties:

1. *Peak flow rate*: Higher peak flow rates indicate that floods occur more frequently and with greater intensity. A higher peak

flow rate necessitates the installation of storm pipes and drains that are larger and more expensive.

2. *Volume*: A higher volume means that there is more energy to scour and erode creek banks, which can result in bank instability, greater sediment deposition in the lower reaches, and the loss of habitat.

3. *Impermeable area*: The impermeable area is characterized by the inability of groundwater to be replenished and the reduction of baseflows, which results in longer and drier periods of time in canals and streams.

4. *Water quality*: Inadequate runoff management could increase contamination and pollution in our waters, which would affect wildlife.

For instance, an increase in impermeable surfaces leads to a significant rise in stormwater runoff compared to the quantity of stormwater that gradually infiltrates the surrounding ecosystem. As a result, the severity of flooding increases. Therefore, it is critical to select management strategies that have proven to be the most effective in reducing the quantity and improving the quality of stormwater runoff to pre-construction levels. The primary purpose of the selected stormwater management practice(s) should be to control stormwater volume while maintaining water quality.

Stormwater management is broadly defined as the planning, maintenance, and regulation of stormwater runoff from rainfall and snowmelt, including collection, storage, and transportation. It aims to safeguard the environment, reduce flooding to protect people and property, reduce mass pollution, promote healthy streams and rivers, and build healthier, more sustainable communities. Local communities gain from effective stormwater management in terms of the environment, social issues, and economic development. Stormwater management that is done properly results in cleaner streams, rivers, and lakes, lower flood risks, lower flood damage costs, and a higher quality of life in the community.

3.1.12 Threat Presented by Insufficient Stormwater Management

There is a correlation between urbanization and an increase in the quantity and decrease in the quality of stormwater runoff that is discharged into our local water sources, such as rivers, creeks, wetlands, and coastal waters. This runoff may contain a significant amount of pollutants that could harm our ecosystem. Furthermore, urbanization results in more impermeable surfaces such as sidewalks, streets, parking lots, rooftops, etc. Impervious surfaces have a major impact on the increase of stormwater runoff by preventing water from infiltrating into the ground. As stormwater runoff moves across impermeable surfaces, it picks up pollutants and carries them to bodies of water like lakes, rivers, and wetlands. These pollutants include fertilizers, soap, oil, dirt, metals, solvents, and feces from pets. In addition to being difficult to control, these pollutants, which are frequently referred to as *nonpoint source pollution*, are primary contributors to water contamination. Inadequate stormwater management not only presents a risk to the environment but also to the economy. Implementing pollution control measures at the source through various stormwater management practices, such as natural vegetation or grass swales, infiltration basins, or detention or retention structures, are typically more cost-effective and simpler than remediation efforts for polluted water bodies. By putting in place stormwater management systems, there will be a decrease in the volume of pollution that originates from stormwater runoff and enters nearby water bodies.

3.1.13 Overcoming Challenges to Improve Stormwater Management for the "New Normal"

Over the course of time, the framework that stormwater managers employ in order to make decisions is regularly subject to modifications. There is the potential for alterations to be made to the policy and legal framework, in addition to the expectations of the community. The implementation of a flexible approach to stormwater management and the awareness of this necessity by those in charge of managing stormwater are both significant. Stormwater managers need to address a wide range of challenges to effectively do their jobs, including municipal finances, dealing with various bureaucracies, developing better management skills, and understanding new technologies. It is rare for a single organization within a catchment (watershed) to be responsible for the management of wastewater and stormwater. The local council, the water authority, the road authority, the rail authority, the catchment management bodies, the state or territory environmental protection authority, the state or territory natural resources management authority, the state or territory planning authority, the national parks management authority, land or property developers, and builders can all play a role. When it comes to management, these organizations typically assign unique roles, tasks, and priorities to each of their departments independently. It is possible that just a small number of these organizations, if any, have stormwater management as one of their core activities.

It can be very difficult to develop a shared vision among the stormwater managers working within a watershed, but it is beneficial for them to try and collaborate on the development of a catchment management plan in order to find solutions to the institutional difficulties that have been presented. The following sections describe some of the challenges that must be overcome in order to improve stormwater management in preparation for the "New Normal."

3.1.13.1 Deficient Management Skills

The transition to a watershed strategy that is more ecologically conscious and integrated has prompted the need for new and broader management abilities. This is because stormwater management is becoming an increasingly multi-disciplinary field. Many municipal councils and other stormwater managers do not possess the technical and administrative expertise to recognize and successfully implement stormwater management plans that strive to achieve multiple objectives. It may be possible to find a solution to this problem through the utilization of continuing education, the employment of consultants, the utilization of other resources within a council (such as environmental and community relations officers), or the sharing of technical resources among stormwater managers.

3.1.13.2 Limitations in Funding

Limitations in funding have led stormwater management authorities to mostly prioritize the installation and upkeep of drainage infrastructure. Securing funding to expand the current stormwater management approach and transition to a more environmentally conscious strategy in response to extreme weather events and changing rainfall patterns may not be easily accomplished through current revenue streams. Structural stormwater management controls can be quite costly, both in terms of initial investment and operating costs. Source controls that are not related to the structure usually have a lower cost. Here are a number of effective strategies that can be utilized to address the funding challenges of stormwater management by municipal authorities, either separately or together:

1. Maximize revenue potential by exploring potential adjustments to rates or fees. When the community is actively involved in the stormwater planning process and has a clear understanding of how their funds are being allocated, you are more likely to receive their support.

2. Grant funding has long been a fundamental component of stormwater management programs. Various grants can be obtained from federal, state, local, and private sources. Grant funding has become a key funding strategy for numerous local stormwater programs.

3. Re-allocate existing funding. Local councils and stormwater programs may have separate budgets related to broader stormwater management. These can include waste management, road maintenance, park maintenance, and community relations. A multidisciplinary approach to stormwater management can utilize resources from these and other sources. It is often possible to combine improved stormwater management practices with existing programs at minimal cost.

4. Special stormwater rates or levies could be determined based on the *generator pays* principle. Landholders who generate larger runoff volumes or pollution loads may be subject to higher fees. This could be determined by analyzing the impermeable surfaces within a specific area of land. Rebates could be offered for the installation of rainwater tanks or measures that decrease the number of impermeable surfaces connected to the drainage system. The allocation of expenses should be fair and in accordance with the law.

5. Additional possibilities could involve the implementation of catchment-based or regional stormwater management programs. This may involve implementing education campaigns or conducting catchment audits of industrial or commercial premises. Developing stormwater management plans collaboratively with all stakeholders can lead to the identification of the most efficient, equitable management practices and decision-making processes that successfully incorporate long-term and short-term economic, environmental, and social factors.

3.1.13.3 Uncertainty

Stormwater management is a complex and challenging field with many uncertainties to consider. These factors encompass a wide range of environmental concerns, such as the release of pollutants from various land uses, the consequences of pollutant loadings on water bodies, the reactions of aquatic ecosystems to shifts in water quality and flow, the impact of modified flow patterns on channel structure, the efficacy of stormwater treatment methods and management strategies, the precision of water quality and hydrological modeling, the difficulties in predicting floods for flood prevention, the expenses associated with reliable and representative water quality monitoring, and the influence of climate and seasonal variations.

There is a wealth of evidence indicating that urbanization and climate change have detrimental effects on aquatic ecosystems. It is important not to allow the current difficulties in measuring these effects to deter us from taking steps to reduce them, and it is crucial to consider this uncertainty when developing a stormwater management planning process. Implement a well-informed and adaptable approach to stormwater management. This may involve executing the most efficient solution to a problem, closely monitoring its effectiveness, and making necessary adjustments. Additionally, it involves promoting the practice of independent monitoring of stormwater treatment devices and openly sharing the findings with other stakeholders. The importance of monitoring also applies to structural solutions and community education programs.

3.1.13.4 Public Involvement

Participation of the public in stormwater management is essential to provide the community with the opportunity to establish *ownership* of both the problem and the solution. It is highly improbable that a strategy that involves the development and application of management solutions in isolation from the community will be successful or sustainable. The process of planning for stormwater management can be difficult to do when it comes to achieving meaningful community consultation. There is a possibility that holding an organized public meeting may not be successful in reaching a substantial portion of the community; a more *boots-on-the-ground* approach may be necessary. The community is likely to be more willing to participate in stormwater planning if the issues are brought to the forefront of public attention. "Adopt-a-drain" programs for businesses and schools as well as community involvement in straightforward stormwater monitoring projects are examples of items that can be helpful in this area. On the other hand, there are several instances in which urban communities have served as the principal source of vision and energy for the process of regaining the qualities of urban waterways with stormwater management authorities providing the technical information in these situations.

3.1.13.5 Urban Development Impacts

The development of urban areas includes an increase in urban density, the expansion of industrial regions, and the failure of aged sewers—all of which have typically led to an increase in the contamination of local and downstream ecosystems by urban runoff. Flooding issues have arisen because of inadequate provisions for stormwater drainage, which have resulted in increased peak flows, increased flow volumes, and altered flow patterns. There is a correlation between increased fertilizer and pollutant loadings and more noticeable changes in the composition and structure of ecosystem types. There should be a reduction in the cumulative loading of pollutants from catchments to levels that are consistent with the sustainability of streams farther down the line. As a base for growth, we should use a coordinated plan that includes catchment-based planning and management, integrated urban land-use planning, integrated water cycle management, and building and managing infrastructure in a way that takes into account social, economic, and environmental values. It is essential to conduct an analysis of the effects that urban development has on the surrounding area through the process of land-use planning. This way, a planning authority can make an informed judgment regarding the advantages and environmental costs of a proposed development.

Green city planning is a collection of activities and methods that are put into place to ensure cleaner air and conserve water and soil for sustained biodiversity. The *Green City Strategy* tries to address urban concerns connected to resource inefficiencies, negative environmental and social costs, and inequalities. It is designed to overcome the challenges that are associated with urbanization. Five priority areas are outlined in this strategy:

1. Ensuring that green growth is incorporated into urban planning and implementation
2. Developing cities that are more resource-efficient and low-carbon in order to support a circular economy
3. Developing decentralized solutions for sanitation and wastewater that better meet the needs of the urban poor and minimize the environmental impact of untreated effluent
4. Managing urban waste through waste-to-resource strategies
5. Developing cities that are more connected and more easily accessible in order to address the difficulties of transportation

and mobility, which in turn encourages the development of healthier cities with higher air quality

3.1.13.6 Sustainable Stormwater Management

Intelligent stormwater solutions should be implemented with a focus on environmental stewardship. Cities are grappling with the challenge of incorporating natural and artificial ecosystems into sustainable development for the well-being of society and future generations as they expand. Experts in the domains of engineering, architecture, and planning are the ones who are able to provide the technical expertise that is necessary to develop, document, design, and put into action solutions to these challenges. When it comes to projects that range from residential and commercial subdivisions to large-scale master-planned communities, we begin by constructing a vision for the project. After that, we employ full-service, quality civil engineering to make that vision a reality. Within the framework of a sustainable stormwater management strategy, both the natural environment and the built environment are managed as integrated components of a watershed. Through the use of this technique, the interdependence that exists between the two environments that are controlled is acknowledged.

The sustainable management of stormwater is an alternative to the traditional approach of managing stormwater through the use of pipes. Consequently, it encourages the collection and conveyance of rainwater from rooftops, parking lots, streets, and other surfaces to penetrate into the ground or gather for reuse, which frequently avoids the need for costly subterranean infrastructure because it allows stormwater to be collected and transported. To slow down and filter water more effectively, sustainable stormwater management takes advantage of natural systems that are completely covered in vegetation. The leaves of the vegetation help increase the amount of precipitation that plants absorb as well as the amount that evaporates. When there is vegetation present, there is a reduction in the number of toxins that are found in urban runoff as well as the volume of stormwater runoff. The process of maintaining sustainable stormwater management typically involves the utilization of both structural and nonstructural devices. Some examples of nonstructural devices include landscaped swales and in-filtration basins. Rain barrels, cisterns, and planters are all examples of structural devices that can improve the environment.

Taking an environmentally friendly approach addresses a number of concerns that are associated with stormwater runoff, including erosion, water contamination, combined sewer overflows, and issues connected to other stormwater runoff-related problems. It is also considerably cheaper than increasing the infrastructure surrounding new buildings and redevelopment projects. Aiming to replicate nature by incorporating stormwater management into the construction of buildings and the development of sites is one way to accomplish the goal of minimizing the adverse effects that urbanization has on our natural resources.

3.2 URBAN DRAINAGE SYSTEMS

Drainage systems are typically classified into two primary types: a combined sewer system (CSS) and a municipal separate storm sewer system (MS4). A CSS handles both stormwater and wastewater (sewage) using a single system, which is then conveyed to a wastewater treatment plant before being released into nearby water bodies. In contrast, an MS4 is engineered to keep stormwater separate from wastewater by redirecting stormwater into a distinct piped system.

While many historical cities and emerging economies still use a CSS to transport urban wastewater, most modern urban stormwater drainage infrastructures are designed as separate systems. CSSs in historical and older cities often contribute to water quality issues, particularly during heavy rainfall events caused by extreme weather, which can overwhelm the combined sewer. In such situations, the wastewater treatment plant releases any excess flow, a mixture of stormwater and sewage, directly into nearby water bodies. This overflow, referred to as a combined sewer overflow (CSO), is implemented to mitigate the risk of flooding in various areas, including residential neighborhoods, basements, commercial establishments, and roadways. Figure 3.9 illustrates a representative urban drainage system.

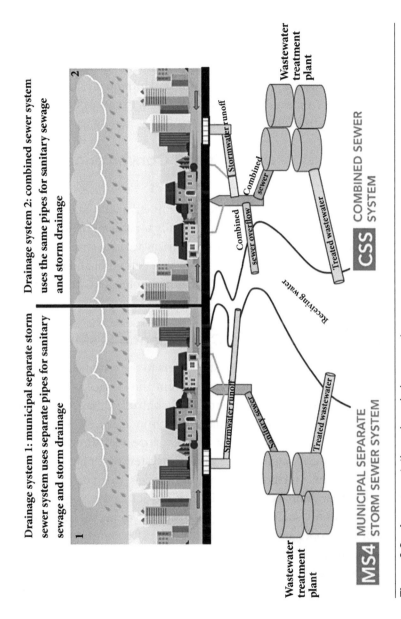

Figure 3.9   A representative urban drainage system

Stormwater runoff management has typically relied on underground systems, including pipes, catch basins, inlets, curbs, gutters, and related infrastructure, and can be classified into two distinct categories: a minor stormwater runoff management system and a major stormwater runoff management system. These systems aim to manage stormwater runoff effectively, especially during severe and infrequent storm events characterized by substantial precipitation. Such events often generate runoff volumes exceeding the minor drainage system's capacity.

3.2.1 Minor Stormwater Runoff Convenience Systems

Minor stormwater runoff convenience systems efficiently handle stormwater runoff of smaller amounts, specifically for storms up to 10-year events. The minor drainage system comprises various structures and components, including pipes, channel swales, detention and retention basins, curbs, gutters, catch basins, inlets, manholes, and other similar components.

3.2.2 Surface Drainage Convenience Systems

A surface drainage convenience system refers to the systematic method of removing excess stormwater runoff from the surface of the land. This process involves the intentional and controlled management of stormwater flow across land surfaces, which includes contouring or grading the terrain to establish a suitable slope that directs surface runoff toward designated surface drainage systems (see Figure 3.10). These systems may include pipes, inlets, channels, catch basins, storm sewers, or other drainage systems. Factors impacting the magnitude of stormwater runoff, such as the length of the flow route, surface roughness, surface slope, and rainfall intensity, must be considered. Each of these elements plays a crucial role in determining the amount of stormwater accumulating on the pavement surface.

Inlets, designed to collect surface stormwater runoff using grate or curb openings, facilitate its transportation to stormwater drainage systems. The design of appropriate grate frames must account for the

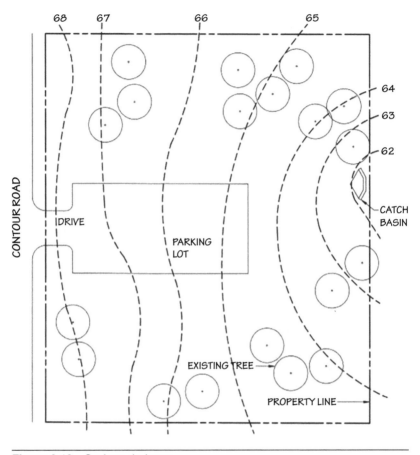

Figure 3.10 Surface drainage

required inlet function, such as traffic, load-bearing, and flow interception capacity. In general, inlets can be divided into three primary categories:

1. *Combined or integrated inlets*: These inlets consist of a configuration where both curb openings and grate inlets are positioned adjacent to each other.
2. *Curb-opening inlets*: Also known as vertical curb openings, these are openings within curbs that facilitate the flow of stormwater runoff into the drainage infrastructure or BMPs.

3. *Grate inlets*: These are openings in gutters equipped with one or more grates, allowing stormwater runoff to enter the inlet through the grate. This distinguishes it from a curb opening. Grate inlets can be used in on-grade situations in combination with curb inlets.

Curb inlets and catch basins (see Figure 3.11) are vital components of underground drainage systems that collect surface stormwater runoff through a piped network. Typically located at low points of a project, such as streets, channels, parking lots, hardscapes, and lawns, catch basins and inlets comprise a grate or curb inlet through which stormwater enters the catch basin and a sump for collecting sediment, debris, floatables, and pollutants. In this context, the terms *curb inlet* and *storm drain inlet* are interchangeable.

Catch basins and inlets come in a wide variety of sizes, shapes, and materials, ranging from basins with a diameter of three inches to grates with a side length of 48 inches. These options include materials such as concrete, brick, cast iron, steel, brass, and plastic. Dome grates are highly recommended for landscape areas where debris accumulation, such as mulch and leaves, may obstruct a flat grate. It is crucial for

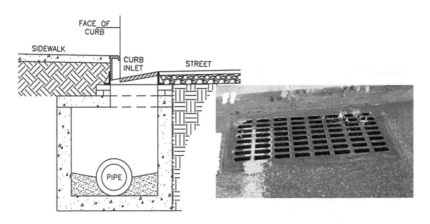

Figure 3.11   Curb inlets and catch basins

designers to thoroughly evaluate drainage requirements before selecting grates and performance functions.

While this book does not cover the design of a grate inlet, the capacity of an inlet depends on its geometry, cross and longitudinal slope, gutter flow and depth, and roughness coefficient. The flow rate, measured in cubic meters per second (m³/sec) or cubic feet per second (cfs), is crucial in determining the capacity of gutter and curb-opening inlets. Grate inlets are often efficient in capturing the entire volume of design or permissible stormwater runoff, including both frontal and lateral flow (see Figure 3.12). However, increased runoff flow velocity, obstruction, or poor design can lead to bypass flow, resulting in only partial interception of stormwater runoff. Therefore, designers must consider multiple factors, including the hydraulic capacity of storm drain inlets, their geometry, and the characteristics of gutter flow.

Linear channel drains (see Figure 3.13) are longitudinal subsurface channels made from precast units, cast-in-place concrete, or grass swales. The top of the channel is flush with the surface, allowing stormwater runoff or sheet flow to drain into grate inlets connected to underground drainage pipes.

Natural surface drain swales, shown in Figure 3.14, are the most cost-effective method of removing unwanted surface water runoff. A typical swale can range from 2 to 60 feet wide and 2 to 60 inches deep.

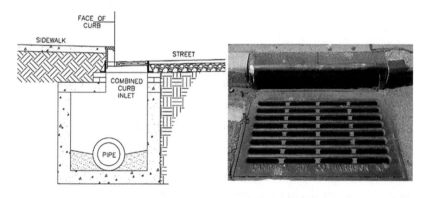

Figure 3.12 Combined curbs and grate inlet

Figure 3.13   Linear channel drains

Figure 3.14   Example of natural surface drain swales

3.2.3 Underground Drainage Convenience Systems

Underground drainage convenience systems (UDCSs) use pipes in minor stormwater drainage systems to transport runoff from roads and other inlets to downstream outfalls, BPMs (structural/nonstructural stormwater controls), or receiving waters. These systems are most effective in medium- to high-density residential and commercial/ industrial developments where open channels and native plant/vege-tated swales are ineffective. They can be made from a wide variety of materials.

3.2.3.1   Polyvinyl Chloride and High-Density Polyethylene

These types of pipes (see Figure 3.15) are popular in drainage applica-tions due to their resistance to collapse and chemical corrosion, as well as their flexibility and durability. High-density polyethylene (HDPE) pipes are more suitable for high-pressure systems than low-cost poly-vinyl chloride (PVC). The ability of HDPE pipes to absorb shock-waves and reduce surges makes them an attractive option for use in underground piping. They are also more resistant to heat and abrasion and can withstand higher pressures in the joints. However, relative to ductile, concrete, or clay pipes, the carbon footprint per linear foot of PVC pipes during transportation and installation is smaller. See Ta-bles 3.1 and 3.2 for the projected lifespans of HDPE and PVC pipes.

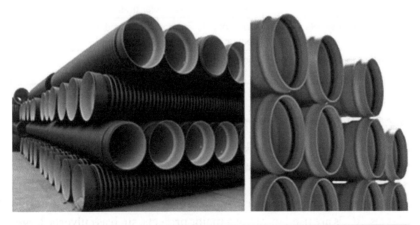

Figure 3.15 Examples of polyvinyl chloride and high-density polyethylene pipes

Table 3.1 High-density polyethylene pipes—pros and cons given the projected lifespan of between 50 and 100 years

Advantages	Disadvantages
Flexible enough to withstand pressure surges and earthquakes	Possesses a high degree of thermal expansion
Easy to rehabilitate with trenchless technology and available in desired lengths, reducing joints	Lacks resistance to acids, chlorinated hydrocarbons, and ketones
Smooth inner surface is abrasion-resistant and inhibits the growth of bacteria and microbes	Can be damaged by ultraviolet light and cannot be used in areas directly exposed to sunlight
Water-resistant, hydrogen-sulfide-corrosion resistant, fused joints prevent leaks and infiltration	

Table 3.2 Polyvinyl chloride pipes—pros and cons given the projected lifespan of between 50 and 70 years

Advantages	Disadvantages
Easy to install and easily adjustable to the required length	Supports less load than high-density polyethylene pipes and reinforced concrete pipes, especially in areas with high vehicular traffic
Low cost and simple to maintain	Good backfill support is required to prevent pipe sections from bending
Resistant to chemical deterioration and corrosion	May warp or distort under extreme heat

3.2.3.2 Reinforced Concrete Pipes

Reinforced concrete pipes (RCPs) (see Figure 3.16) can last for decades if properly installed. Despite their vulnerability to acidic chemicals or soil degradation, RCPs can operate at high pressures and resist root and soil intrusion. Available in different wall thicknesses and design classes, RCPs are used in various major projects, such as culverts, large-scale irrigation, and sanitary and storm sewers, due to their proven lifespan (see Table 3.3), which often exceeds 100 years.

Figure 3.16   Examples of reinforced concrete pipes

Table 3.3   Reinforced concrete pipes—pros and cons given the projected lifespan of between 75 and 100 years

Advantages	Disadvantages
Superior durability	Proper handling of the installation process is crucial to prevent potential damage
High-pressure resistant and not easily broken	Requires heavy-duty machinery for installation
Prevents the infiltration of roots or soil	The roughness of its inside surface results in sediment accumulation

3.2.3.3 Vitrified Clay Pipes and Fittings

Vitrified clay pipes (VCPs) are commonly used in drainage and sewage construction projects of various sizes and complexity levels. Known for their exceptional ecological compatibility and impressive durability, especially in corrosion-prone environments, VCPs are a favored option for advancing drainage and sewerage infrastructure projects, resulting in reduced construction timelines. While they are cheaper and more chemically resistant than concrete pipes, they are also more prone to root intrusion and cracking than concrete or HDPE pipes. Despite being able to withstand abrasion, VCPs can become brittle, leading to cracks that allow root intrusion into the waterline. In some areas, it is better to use metal pipes instead of VCPs. While clay can withstand aggressive cleaning, the internal pipe pressure is lower than that of metal pipes, making this type of plumbing material unsuitable for areas with higher overall water pressure (see Table 3.4).

Table 3.4　Vitrified clay pipes—pros and cons given their projected indefinite lifespan

Advantages	Disadvantages
A more economically viable alternative	A lower degree of pressure tolerance than metal pipes
Highly resistant to chemical and acidic erosion	Requires heavy-duty machinery for installation and has a limited length
Possesses the most extensive lifespan among all presently used options; superior joint designs contribute to their overall strength and reliability	Lacks the flexible properties necessary to withstand excessive forces without fracturing; crucial for maintaining hydraulic integrity
Rehabilitation can be carried out with enhanced convenience, including trenchless rehabilitation technology	Increased susceptibility to brittleness raises the likelihood of cracking and makes it more vulnerable to root intrusion

3.2.4 Major Drainage Systems

A well-designed and highly efficient stormwater drainage system, often referred to as a major drainage system, is crucial for managing large volumes of intense storm events that exceed the capacity of minor storm drainage systems. High-intensity storm events, such as those occurring every 25, 50, or 100 years, can produce substantial amounts of stormwater runoff, posing a significant risk of widespread flooding. Major drainage systems play a significant role in safeguarding against negative outcomes associated with these severe weather conditions, including protecting human lives and property. The primary stormwater drainage system comprises many different components, dependent on hydrology and hydraulic analysis.

3.2.5 Overland Stormwater Runoff Discharge Controls

Overland stormwater runoff discharge refers to the movement of stormwater during and after a rainfall event. The flow rate of stormwater is typically quantified in cubic meters per second or cubic feet per second. To counteract the potential negative impact of increased peak overland stormwater runoff or volume entering receiving waters, regulators often mandate the implementation of stormwater runoff discharge controls. Various techniques, including the use of detention and retention basins, larger storm drainage pipes, swales, channels, and other storage facilities with outlet controls, can effectively manage excessive stormwater runoff. These methods are crucial for mitigating the effects of heavy rainfall, managing stormwater runoff efficiently, and alleviating downstream flooding problems.

3.2.6 Overland Stormwater Quality Controls

Stormwater runoff poses a significant concern in urban environments due to its potential to transport a variety of pollutants and contaminants into aquatic ecosystems. The strategic development of BMPs is essential to optimize pollutant mitigation while ensuring cost-effectiveness. This strategy involves a systematic approach to collecting and treating stormwater runoff, primarily aimed at reducing the introduction of pollutants into water bodies. Efficient stormwater management strategies can help preserve the quality of our water resources. Implementing BMPs at strategic locations within a watershed is crucial to effectively reduce pollutants.

BMPs are primarily classified into structural and nonstructural methodologies. Structural BMPs involve the use of tangible structures such as rain gardens, retention and infiltration-based structures, and constructed wetlands, designed to manage stormwater runoff and facilitate natural infiltration processes. Conversely, nonstructural BMPs focus on restoring the natural environment and mitigating the effects of impervious surfaces. Measures such as implementing riparian stream buffers—vegetated areas along the banks of streams or rivers—can achieve this goal. Enhancing these buffers can effectively protect the surrounding ecosystem, leading to significant improvements in water quality. Disconnecting impervious areas, such as driveways or parking lots, from the stormwater system can mitigate runoff and prevent the contamination of water bodies with pollutants.

Both structural and nonstructural BMPs are vital components in managing stormwater effectively and safeguarding our valuable natural resources. By incorporating these strategies, we can address the adverse effects of urban development and promote sustainable water management. In the context of enhancing water quality, retrofitting existing BMPs can be a financially prudent strategy to improve treatment capabilities. This approach offers a pragmatic and effective solution for improving water quality without incurring substantial additional expenses.

3.3 BASICS OF STORMWATER CONVENIENCE SYSTEM DESIGN

The design of stormwater drainage is crucial in the planning and implementation of both site-specific and broader stormwater management strategies. The primary objective of effective drainage design is to ensure compatibility with preexisting drainage patterns and to minimize disruption. It aims to manage and mitigate property, structure, and roadway flooding during flood events and to minimize potential adverse environmental consequences associated with stormwater runoff. The design of stormwater collection systems must provide sufficient surface drainage while addressing additional stormwater management objectives, including water quality enhancement, streambank channel protection, habitat preservation, and groundwater recharge promotion. This section summarizes the key principles of hydrologic analysis and stormwater management system design. For further information, readers are advised to refer to relevant undergraduate courses, particularly those related to hydrology and hydraulic analysis.

3.3.1 The Project Site Development Process

The procedure for designing the project site development and stormwater convenience system may vary based on project specifics. The process typically involves site feasibility studies, site design, planning and layout, earthwork and grading, parking design, streetscape design (including highways and roads), floodplain analysis, erosion control, sewage facility planning, stormwater analysis, management and design, and utility coordination.

3.3.2 Site Survey and Analysis

Analyzing a site survey is a vital step in assessing the feasibility of construction projects for specific locations. This process involves gathering information about the site and analyzing it to determine the suitability of alternative drainage features for the intended project. Data pertaining to a range of factors, including topography, soil conditions, hydrology, existing utility infrastructures, and potential ecosystem and environmental impacts, are collected and examined. This comprehensive analysis provides a thorough understanding of the project site. The scope and complexity of the site survey and analysis should align with the significance, size, and complexity of the project since these data significantly impact the overall decision-making process.

For a drainage system design project, it is necessary to collect hydrologic data, including precipitation patterns, nearby water bodies, groundwater levels, and the overall water cycle in the region. Understanding these factors allows project planners to better assess potential challenges and opportunities when selecting alternative conveyance systems. For instance, if the project area experiences high rainfall levels or has a high water table, alternative systems that can effectively manage and mitigate these conditions may be necessary. It is also important to consider the locations of hydraulic features, such as reservoirs, water projects, and designated or regulated floodplain areas because these features can significantly influence flood patterns and should be considered in any comprehensive survey or analysis.

In the context of a project site, various factors contribute to its overall layout and functionality. These factors include site boundaries, site areas, utilities, contours, dimensions, site features, climate, and legal information. Site boundaries, the defined limits of the project site, must be clearly identified and demarcated to ensure the project remains within the designated area. Site areas, the different sections or zones within the project site, may vary in terms of their purpose, such as residential, commercial, or recreational zones. Understanding the site areas aids in effective planning and resource allocation.

Collecting existing utility service information, including electricity, gas, water, sewer, and telephone, is vital in any project site analysis. Identifying the locations of these utilities, both within and near the site, including the horizontal and vertical distances, depths, and materials used for these utilities, is crucial for proper planning and coordination. Contour information provides insight into the natural topography of the site, helping project planners determine the best ways to utilize the land and accommodate any necessary modifications or adjustments. The physical measurements of the site, pertaining to its overall size and shape, are crucial for efficient design and successful project implementation. Site features, the distinct characteristics or elements present within the project site, may encompass existing structures, vegetation, water bodies, or any other notable features that necessitate consideration throughout the planning and implementation stages. Conducting a comprehensive analysis and investigation during the early stages of the site survey can help identify factors that may negatively impact the site or proposed design, as well as conditions that exert a more favorable influence.

3.3.3 Hydrology and Hydraulic Analysis

The analysis of stormwater runoff hydrology and hydraulics is paramount in the design of drainage infrastructure. It is crucial to consider various factors that could potentially impact the intended results. These factors include the peak rate, volume, and time distribution of stormwater runoff. For instance, a particular design may require a peak flow rate, defined as the maximum flow rate at which stormwater runoff can flow within a specific time frame. In contrast, an alternative design may necessitate the use of a runoff hydrograph, a visual representation used to estimate the volume of stormwater runoff. These factors underscore the multifaceted aspects that must be considered when designing different drainage infrastructure systems. The peak flow rate is of utmost importance in the design of various infrastructure components, including pipes, bridges, culverts, and channels.

The development of a runoff hydrograph is crucial in the design of various infrastructure components, including detention storage, pumping stations, and large or complex storage systems. The hydrograph aids in understanding and analyzing stormwater flow patterns during storm events. Through a comprehensive analysis of the runoff hydrograph, engineers and drainage professionals can make informed recommendations regarding the optimal drainage system for a given scenario. This measure ensures maximum efficiency in stormwater management and risk reduction. It is also vital to be aware of potential errors that may arise during hydrology and hydraulics analysis. These errors can have substantial implications for the estimated structure. An undersized structure can lead to increased drainage problems, while an oversized structure can result in unnecessary costs. Therefore, accurate estimates are crucial to avoid these undesirable outcomes.

3.3.4 Design Flood Frequency (Return Period-Years)

Design flood frequency refers to the expected occurrence rate of a specific rainfall event over a given period of time (years). A typical hydraulic facility's design flood frequency is influenced by several factors, including the type, size, and geographical location of the drainage structure. There are two widely accepted methods for flood-frequency selection in designing a drainage system. The first is a predefined design flood-frequency method that adheres to established design guidelines, standards, and policy. The second is a contemporary design concept that conducts an economic analysis. The use of traditional predefined design flood frequencies implies that a thorough risk assessment was performed while defining the design flood-frequency standard. Contemporary design concepts advocate for the evaluation of a variety of peak flows and the construction of a design flood frequency that best meets the individual site characteristics and related risks. Both options are valid and can be used separately or in conjunction, depending on overall conditions or specific limits. For further information on frequency analysis, refer to Chapter 2, Section 2.7, of this book.

3.4 TIME OF CONCENTRATION AND TRAVEL TIME

The concepts of time of concentration and travel time are significant in the field of hydrology. *Time of concentration* refers to the duration it takes for a drop of stormwater to travel from the farthest point in the drainage area, watershed, or subwatershed, determined by the highest hydraulic distance, to the specific location under investigation or analysis. The *hydraulic length*, also known as the *flow path*, pertains to the pathway followed by stormwater runoff. *Travel time* in hydrology refers to the time required for a drop of stormwater to travel from one site within a subwatershed to another predefined location within the total watershed. The travel time delineates the spatial positioning of the subwatershed in relation to the focal site being studied within the total watershed. Thus, the time of concentration is the total duration required for a drop of stormwater to travel across distinct flow segments in a given watershed, plus the individual travel time values for each successive flow region. Stormwater runoff experiences various flow types or flow regimes throughout its course, such as overland or sheet flow, shallow concentrated flow (SCF), channel flow, and pressurized pipe flow. The following sections provide a brief discussion of the various flow regimes:

- *Overland flow*: Overland flow, also known as sheet flow, refers to the shallow movement of stormwater runoff, typically less than 2.54 centimeters deep, across relatively flat surfaces. The duration of overland flow can vary depending on specific field conditions. To accurately measure overland flow length, a thorough topographic survey is essential. This survey identifies various surface features within the project area and assesses the duration of overland flow. Such information is vital for professionals designing drainage infrastructure because it validates elevation changes, terrain levels, and other relevant factors. A comprehensive analysis of the terrain and meticulous study of flow patterns ensure the accuracy and reliability of the findings. Topographic surveys provide a detailed assessment of an area's physical characteristics, enabling informed assessments regarding overland flow duration and design.
- *SCF*: At the end of overland flow, water begins to collect in small gullies and grassy depressions. This flow condition, lacking the properties of a well-defined channel flow but not resembling a thin sheet of water, is best described as an SCF transitional condition that occurs between sheet flow and channel flow. This transition happens when overland flow converges, forming tiny channels like swales, rills, and gullies. These runoff conveyance structures are crucial for successfully diverting stormwater runoff into the urban drainage system. Shallow concentrated flow can occur in various situations, often seen in metropolitan settings featuring small paved or unpaved ditches, roadway curbs, and gutters designed to manage stormwater runoff.
- *Channel flow*: In urban drainage infrastructure, channel flow refers to a permanent, well-defined channel cross-section or

waterway designed to convey concentrated surface runoff without causing erosion or chronic flooding risk. Channel flow conveyance systems can take various forms, such as partially full pipes, vegetation or riprap-lined channels, gullies, ditches, or other types of water conveyance systems, whether natural or man-made.

- *Pressurized pipe flow:* this term refers to storm drainage pipes that are functioning at or near capacity (flowing full).

3.4.1 Methods for Calculating Time of Concentration

Determining the *time of concentration* is a critical aspect that involves multiple factors. These factors include the size and configuration of the drainage basin, the area's topography, the type of land use and level of urbanization, soil characteristics, the presence of natural channels, the existence of man-made drainage systems, and the impact of storage mechanisms such as wetlands, ponds, and reservoirs. When calculating the time of concentration, it is essential to consider all these factors.

It is a common practice to first determine the time of concentration and then use it to calculate storm values for various frequencies. However, it is important to note that the duration of concentration may vary across different storms. The determination of time of concentration often involves subjectivity and heavily relies on the engineer's expertise and judgment. Using a range of methodologies is crucial to ensure the accuracy and reliability of the results, verify the results effectively, and validate their coherence.

Applying various methodologies for determining the time of concentration is a valuable quality control measure that helps minimize errors and enhances the overall reliability of the results. In hydrograph analysis, the time of concentration refers to the duration between the cessation of intense rainfall and the specific point on the descending segment of the dimensionless unit hydrograph, commonly known as the point of inflection, where the recession curve begins. The analysis of time of concentration for the design of urban drainage infrastructure includes the computation of the following:

1. *Lag time*: This is a crucial hydrologic factor in the creation of a unit hydrograph. It refers to the duration between the beginning of runoff from a rainfall event in a watershed and the point at which the runoff reaches its peak. In hydrograph analysis, lag refers to the time duration between the midpoint of the surplus rainfall and the maximum rate of runoff.

2. *Inlet time*: This term specifically refers to the duration required for stormwater runoff to travel from its farthest point to a predetermined inlet location. The estimation of inlet time can be derived by combining the travel time of overland/sheet flow with the travel time of SCF. It is uncommon for channel flow to be part of the calculation of inlet time.

3. *Flow travel time*: In hydrology, travel time refers to the time it takes for a drop of stormwater to transit between two specific locations in a given watershed. Travel time is calculated using the ratio of flow length to flow velocity. This parameter is essential for understanding the dynamics of stormwater flow and its movement within a given environment. By determining its travel time, design professionals can gain insights into the speed and efficiency of stormwater transport, which can have significant implications for various applications such as flood predictions, water resource management, and pollutant dispersion analysis.

Table 3.5 includes the most common techniques employed in the determination of time of concentration. However, further exploration of this topic is beyond the scope of this textbook. Readers are strongly recommended to supplement their understanding by consulting other scholarly resources dedicated to the same subject matter.

Table 3.5 Commonly used methods to calculate the time of concentration

Name	Formula	Description
NRCS Curve Number (U.S.D.A. NRCS 1986)	$$T_c = \frac{(L)^{0.8}\left(\frac{100}{CN}-9\right)^{0.7}}{441Y^{0.5}}$$ Where: T_C = Time of concentration, h L = Length of mainstream to farthest divide, m Y = Average watershed slope, % CN = NRCS curve number	This method was developed to analyze a small rural watershed with natural homogeneity, uniform curve number, area ranging from 1 to 800 hectares, average slope from 5% to 64%, and flow length from 60 meters to 8 kilometers or 97 feet to 5 miles
Kinematic Wave Equation HEC No. 12 (Johnson 1984)	$$T_O = \frac{6.92L^{0.6}n^{0.6}}{(C*I)^{0.4}S^{0.3}}$$ Where: T_O = Time of overland flow, min L = Overland flow length, m n = Manning roughness coefficient C = Runoff coefficient I = Rainfall rate, mm/h S = Average slope of the overland area, decimal	The most practical equation for determining the time of concentration for overland flow
Manning's Kinematic Solution	$$T_O = \frac{0.091(nL)^{0.8}}{(P_2)^{0.5}S^{0.4}}$$ Where: T_O = Overland flow time, h n = Manning's roughness coefficient L = Flow length, m P_2 = 2-year, 24-h rainfall, mm (from TP-40) S = slope of hydraulic grade line (land slope), decimal	Recommended for overland or sheet flow of less than 90 meters or 255 feet

Continued

Table 3.5 *continued*

Name	Formula	Description
Federal Aviation Administration Method (FAA 1970)	$T_O = \dfrac{(1.1-C)L^{0.5}}{144S^{0.3}}$ Where: T_O = Overland flow travel time, min L = Overland flow path length, m S = Slope of overland flow path, decimal C = Rational method runoff coefficient	This equation is particularly applicable to situations where the drainage area is relatively small and the surfaces within it exhibit a high degree of uniformity.
Kerby's Equation	$T_O = K(LNS^{-0.5})^{0.467}$ Where: T_O = Time of overland flow, min K = 1.44 L = Length of flow, m N = Retardance roughness coefficient S = Average slope of overland flow, decimal	The Kerby approach proves to be advantageous in small watersheds where the contribution of overland flow significantly influences the calculation of total travel time. The Kerby and Kirpich formulas may not be suitable for watersheds characterized by a gradual topographic slope. Overland flow is calculated using the Kerby equation, while channel flow is calculated using the Kirpich equation.

Continued

Table 3.5 *continued*

Name	Formula	Description
Kirpich's Equation	$T_C = \left[\dfrac{0.948L^3}{H}\right]^{0.385}$ Where: T_C = time of concentration, h L = Length of the longest waterway from the point in question to the basin divide, km H = Difference in elevation between the point in question and the basin divide (omitting drops due to gully overfills, waterfalls, etc.), m	This equation is appropriate for analyzing watersheds in natural, rural basins with well-defined channels. It is also useful for assessing overland flow on exposed soil surfaces or within roadside channels. The applicability of Kerby's and Kirpich's formulas may be limited to watersheds with a gradual topographic slope. Overland flow is calculated using the Kerby equation, while channel flow is calculated using the Kirpich equation.
Total Time of Concentration	$T_C = T_o + T_t$ Where: T_C = Total time of concentration T_o = Overland flow time T_t = Travel time = $T_t = \dfrac{L}{60V}$ L = Flow length, m V = Average velocity, m/s	

3.4.1.1 Most Common Errors in Calculating T_C

Two common errors should be avoided when calculating the time of concentration, T_C:

1. Failing to properly delineate a watershed by considering various land cover types throughout the drainage area to estimate different time of concentration values. This is necessary to accurately determine the design flow required for a specific application.
2. Neglecting to consider the appropriate path length that corresponds to the specific characteristics of the study area. When conducting an analysis of a study area's cover, slope, and other relevant factors, it is crucial to precisely and accurately consider the overland flow path length.

3.4.2 Methods for Calculating Stormwater Flow Velocity

Understanding the dynamics of stormwater runoff within a watershed is of utmost importance. The movement may exhibit as sheet flow, SCF, open channel flow, or a combination thereof. These different flow patterns significantly influence the distribution of stormwater runoff and its subsequent impacts on the surrounding ecosystem. Analyzing the velocity of water flow and rate enables experts to develop efficient stormwater management tactics. Various techniques exist for determining stormwater flow velocity, valuable for both determining flow velocity and calculating the time of concentration.

3.4.2.1 Natural Resources Conservation Service (NRCS) Upland Method

This technique can be used to calculate flow velocity. It establishes a correlation between the watershed's slope and surface and the flow velocity. The TR-55 method (1986) employed grassed waterway relations to compute the SCF travel time for unpaved and paved areas using Equations 3.1 and 3.2:

Unpaved $\qquad$ $V = 4.9178S^{0.5}$, $\qquad$ (3.1)

Paved $\qquad$ $V = 6.1961S^{0.5}$, $\qquad$ (3.2)

where V is the average velocity (m/s) and S is the watercourse slope or hydraulic slope.

3.4.2.2 Manning's Equation

Manning's equation is a well-established empirical equation utilized in water resources. It is a function of flow velocity, cross-sectional area, hydraulic radius, roughness or friction factor, and slope. The formula has gained significant importance in the field of water resources due to its versatility and extensive application. Manning's equation can perform a range of calculations pertaining to fluid dynamics, including determining the flow rate in an open channel, computing the friction losses that occur within a channel, estimating the capacity of a pipe, evaluating the performance of an area-velocity flow meter, and providing various other potential applications. Manning's equation (Equation 3.3) is mathematically expressed as follows:

$$V = \frac{KS^{0.5}R^{0.67}}{n},$$ (3.3)

where V is the mean velocity of flow (m/s), n is Manning's roughness coefficient (unitless), R is the hydraulic radius (m) ($R = \frac{Area}{wetted\ perimeter}$); (see Figure 3.17), S is the slope of the hydraulic grade line (decimal, not in percentage), and K is the unit conversion factor: 1.49 for U.S. units, and 1 for metric units.

To maintain optimal performance of the urban drainage system, storm drain infrastructures must be appropriately sized to support the pipe flow during ordinary events. It is critical to ensure that all drainage systems function at or close to their maximum capacity for the selected design discharges and frequencies. Manning's equation is commonly used in water resource engineering to accurately determine the velocity of flow in circular pipes. It is especially valuable in scenarios where the flow is either full or nonpressurized (see Equation 3.4):

Channel Type	Cross Section Area (A)	Wetted Perimeter (p)	Hydraulic Radius (R)
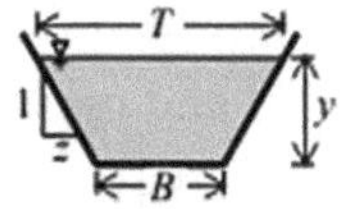 (Trapezoidal diagram)	Trapezoidal Cross Section		
	$By + Zy^2$	$B + 2y\sqrt{Z^2 + 1}$	$\dfrac{By + Zy^2}{B + 2Y\sqrt{Z^2 + 1}}$
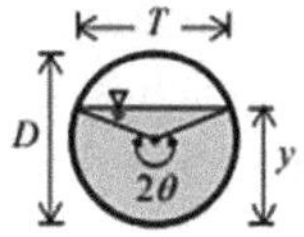 (Circular diagram)	Circular Cross Section		
	$\dfrac{D^2}{8}(2\theta - sin2\theta)$	θD	$\dfrac{1}{4}\left|1 - \dfrac{sin\theta}{\theta}\right|D$
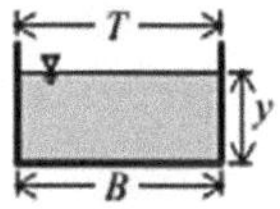 (Rectangular diagram)	Rectangular Cross Section		
	By	$B + 2y$	$\dfrac{By}{|B + 2y|}$
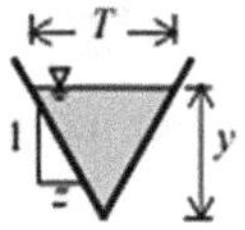 (Triangular diagram)	Triangular Cross Section		
	Zy^2	$2y\sqrt{Z^2 + 1}$	$\dfrac{Zy}{2\sqrt{Z^2 + 1}}$

Figure 3.17 Hydraulic radius, a wetted perimeter of a selected cross section used in urban drainage systems

$$V = \frac{0.397\, S^{0.5} D^{0.67}}{n}, \qquad (3.4)$$

where D is the diameter of a circular pipe (m).

Pipe flow charts are valuable tools for calculating the velocity of fluid flow in pipes. They apply to full and partially full flow conditions. One can obtain accurate velocity values for various flow scenarios by utilizing pipe flow charts. For a storm drainage system where the pipes operate under pressure flow, it is vital to utilize the continuity equation (Equation 3.5) to calculate the velocity:

$$V = Q/A, \qquad (3.5)$$

where V is the mean velocity of flow (m/s), Q is the discharge in pipe (m³/s), and A is the area of pipe (m²).

3.4.2.3 Triangular Gutter Flow

The flow velocity within a gutter significantly influences the analysis of water spreading onto the adjacent parking lane or traveling way. This equation takes into account the gutter's longitudinal and cross-slope, including the spread of stormwater. The average velocity in a triangular channel, characterized by uniform inflow per length and zero flow at the upstream end, is determined when the spread reaches 65% of its maximum value. Equation 3.6 can be used to calculate the flow velocity in a triangular gutter section:

$$V = \frac{0.7575 S^{0.5}(S_x)^{0.67} T^{0.67}}{n}, \qquad (3.6)$$

where V is the flow velocity in gutter (m/s), n is Manning's roughness coefficient for sheet flow, S is the longitudinal slope (decimal), S_x is the gutter cross-slope (decimal), and T is the water spread (m).

3.5 CALCULATING PEAK DISCHARGE RATE AND HYDROGRAPH

A thorough understanding of stormwater runoff volume, peak rate, and hydrograph is crucial for effective stormwater runoff management and environmental impact minimization, as well as for the design of efficient drainage infrastructure. The evaluation of stormwater runoff typically involves established methodologies such as the rational or modified rational methods and the Soil Conservation Service (SCS) or NRCS methods. These assist in determining stormwater peak flow and runoff volume. Despite the development of various complex methodologies for rainfall-runoff modeling, these three methods remain widely used for urban drainage calculations due to their ability to simplify the modeling processes. These methodologies are commonly applied in drainage infrastructure design, including flood forecasting, storm sewers, pipes, channels, culverts, detention and retention basins, and various other elements. Hydrological engineering methods can be classified based on their ability to accurately simulate flows from various land-use drainage areas and generate output suitable for designing a wide range of drainage infrastructure. The following sections provide summaries of these three commonly used methods.

3.5.1 Rational Method

The rational method is widely used in hydrology to determine peak stormwater runoff discharge in both rural and urban drainage areas. Its user-friendly nature has contributed to its significant popularity and continued use. Equation 3.7, the *rational formula*, is based on three key factors: the runoff coefficient, the design rainfall rate, and the drainage area under study. These factors are used to estimate the peak runoff for the selected design storm:

$$Q = 0.00278CIA, \quad (3.7)$$

where Q is the peak stormwater runoff rate (m^3/s), I is the design rainfall intensity (mm/h) for a selected frequency of occurrence or return period, A is the drainage area (ha; rural area of up to 80 ha and urban area of up to 40 ha), and C is the runoff coefficient (unitless). The value for C can be expressed via Equation 3.8 as follows:

$$C = \beta C_1, \quad (3.8)$$

where C_1 is the runoff coefficient representing a ratio of runoff to rainfall and β is the runoff coefficient adjustment factors (see Table 3.6).

Table 3.6   Runoff coefficient adjustment factors

Recurrence Interval/Design Storm	Runoff Coefficient Adjustment Factor (β)
2- to 10-year	1.0
25-year	1.1
50-year	1.2
100-year	1.25

3.5.1.1 Runoff Coefficient

Determining the value of the runoff coefficient, denoted as C, typically involves a comprehensive assessment of multiple parameters within a given drainage basin. The chosen value must be suitable, considering variables such as storm conditions, infiltration rates, antecedent moisture levels, soil types, land-use patterns, surface depressions, ground cover, and slopes. It is recommended to adjust the chosen C value proportionally as the gradient of the drainage basin increases. This recommendation is based on the relationship between slope and velocity. A steeper slope leads to higher velocities in both overland and channel flow, reducing the capacity for water to infiltrate into the ground surface. Consequently, a larger proportion of the precipitation will convert into surface runoff within the specified drainage region. To calculate the composite coefficient for a drainage area that consists of subareas with varying runoff coefficients, divide the total area by the sum of the products obtained by multiplying each subarea's coefficient with its corresponding area. Equation 3.9 provides the formal mathematical expression:

$$C = \frac{\sum A_1 C_1 + A_2 C_2 \dots . A_n C_n}{\sum A_1 + A_2 \dots . A_n}. \tag{3.9}$$

Readers should consult specialized undergraduate textbooks for a more comprehensive understanding of runoff coefficients for various combinations of ground cover and slope since this textbook does not delve in-depth into this topic.

3.5.1.2 Design Rainfall Intensity

Rainfall intensity, denoted as I, is a measure of the average rate of rainfall, expressed in millimeters per hour. It is related to the duration of rainfall and the recurrence interval of a design storm. Once the return period for the design has been chosen and the time of concentration (T_C) for the drainage area has been calculated, the rainfall intensity can be determined using Equation 3.6 to calculate the peak flow in the rational method. Alternatively, the rainfall intensity can be obtained by referring to the rainfall intensity-duration-frequency (IDF) curves specific to the design frequency. However, it is necessary to first calculate the time of concentration for the study area under consideration (see Equation 3.10):

$$I_f = \frac{B}{[T_C + D]^E},\qquad(3.10)$$

where I_f is the rainfall intensity (2, 5, 10, 25, 50, and 100 year) in inches/hour, T_C is the drainage area time of concentration in minutes, and B, D, and E factors are derived from the National Oceanic and Atmospheric Administration:

- *Rational method runoff hydrograph*: A hydrograph is a graphical representation that illustrates the temporal changes in stormwater runoff at a designated point within a hydrological basin. The integral of the hydrograph represents the summation of the areas, indicating the total volume of runoff. During a period of sustained precipitation, the peak rate of stormwater runoff is observed when all the subdrainage areas within the overall drainage area are actively contributing to the specific location under investigation. This event occurs at a particular time (t) which corresponds to the time of concentration, represented as T_C. Based on the specific design of the rational method for predicting maximum peak discharge, it can be inferred that ongoing rainfall does not have any impact on increasing the peak rate of discharge. Therefore, the rational hydrograph can be generated using a rational, modified, or user-defined ratio. To calculate the peak flow using the rational formula, it is necessary to use a pre-established IDF curve and the time of concentration. All methods require these inputs. Figure 3.18 provides a visual representation of a rational-method runoff hydrograph.

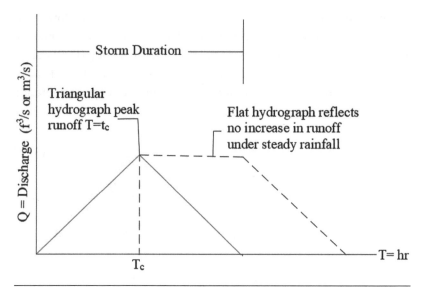

Figure 3.18   Rational method runoff hydrograph

The rational method hydrograph is suitable for estimating peak stormwater runoff rates. However, it lacks precision in calculating the total runoff volume, which is determined by the area under the triangular or trapezoidal hydrograph. This issue arises because the rational method is designed to calculate peak stormwater runoff rates when the entire drainage area is actively contributing. It is possible to generate a primary runoff hydrograph that exhibits a triangular or trapezoidal shape (see Figure 3.17) by establishing a predetermined storm duration known as *time of concentration* that is equal to or longer than the actual duration. Notably, the key variables are not directly associated with precipitation or land-use patterns, except for the intensity of the maximum rainfall return frequency.

3.5.2 Modified Rational Method—Flow Routing

The modified rational method is a hydrological analysis technique used to calculate stormwater runoff rates at specific points within basins that include multiple subbasins. This method generates triangular or trapezoidal hydrographs to determine stormwater storage volume. It employs a methodology akin to the rational method, but it uses a predetermined rainfall duration.

The rational method calculates the peak stormwater runoff flow rate from the combined contribution of the entire watershed at a specific time duration, known as the time of concentration. It does not account for the effects of storms that last beyond this time. In contrast, the *modified rational method* includes storm events with durations that exceed the time of concentration in multiple subbasins. This can result in a variable peak discharge rate, which can be either smaller or larger. However, it will increase the stormwater runoff volume due to the extended rainfall duration. It is important to note that the storm event duration that results in the highest volume of stormwater runoff does not always coincide with the period of maximum peak stormwater runoff discharge rate. Figure 3.19 presents a compilation of hydrographs representing storms of different durations.

It is crucial to recognize that the modified rational method shares the limitations of the rational method. Due to its assumption of uniform storm intensity and disregard for soil conditions, this method provides an approximation rather than an accurate hydrograph. Consequently, it is recommended to limit the application of the modified rational method to drainage basins of five acres or less that do not involve any off-site discharge. The primary difference between the rational method and the modified rational method lies in the assumed shape of the resulting runoff hydrograph.

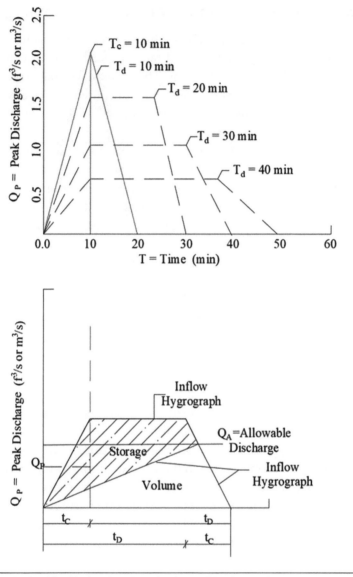

Figure 3.19 Modified rational method runoff hydrographs

3.5.3 NRCS or SCS Unit Hydrograph Method

In hydrograph analysis, the unit hydrograph methods of the NRCS, formerly known as the SCS, are commonly employed to compute stormwater runoff rates. Key parameters for the NRCS method include the drainage area, runoff coefficient, time of concentration, and rainfall intensity. These parameters are essential for accurately calculating the stormwater runoff rate using NRCS unit hydrograph methods. The analysis considers variables such as the spatial and temporal distribution of rainfall, initial losses from interception and depression storage, and the soil's antecedent conditions. Once the excess precipitation is determined, the direct runoff for a given storm, whether observed or simulated, can be computed by subtracting the amount of rain that infiltrates the ground and other losses from the total rainfall. For more information, users are advised to consult Section 4 of the NRCS National Engineering Handbook.

The NRCS method encompasses both the unit hydrograph and the dimensionless unit hydrograph. A unit hydrograph is a graphical representation that depicts the temporal distribution of flow resulting from one millimeter of volume of direct runoff occurring over a specific watershed within a given period. It represents the unique characteristics of the watershed, not the final stormwater runoff hydrograph. Conversely, a dimensionless unit hydrograph is a composite hydrograph that amalgamates multiple unit hydrographs. The properties of the dimensionless hydrograph depend on the dimensions, configuration, and gradient of the tributary drainage area. Two key factors influence the shape of the dimensionless hydrograph: basin lag and peak discharge from a single rainfall event. *Basin lag* refers to the time delay between the centroid of rainfall surplus and the hydrograph's peak. Factors such as natural terrain, gentle gradients, a larger drainage basin, or a parallel drainage system can extend the lag time and reduce the peak stormwater runoff rate.

In contrast, a drainage basin with steep slopes, a smaller size, and an efficient drainage network tends to decrease lag time and increase the peak stormwater runoff rate. The lag time is the duration it takes for stormwater runoff to travel from the point of rainfall to a specific downstream location. The peak stormwater runoff rate is the maximum flow rate or volume of water observed during a given period.

Once established, the unit hydrograph becomes a valuable tool for calculating the runoff hydrograph resulting from various design storms. Hydrologists and engineers can use the unit hydrograph to accurately predict the flow of stormwater runoff in a watershed during different storm events, which is crucial for designing effective stormwater management systems and ensuring the proper functioning of drainage infrastructure. The calculation of the peak discharge for the unit hydrograph, a critical step in hydrological analysis, involves a specific methodology and several key factors. This process allows for the accurate determination of the maximum flow rate that a river or stream will experience during a given storm event, which is vital for applications such as flood forecasting and the design of hydraulic structures.

3.5.3.1 NRCS Equations and Concepts

The NRCS method employs the following equations and basic concepts. The peak discharge for the unit hydrograph is calculated using Equation 3.11:

$$Q = \frac{484 C_A P_X}{T_P},\qquad(3.11)$$

where Q is the peak stormwater runoff flow (cm/s, cf/s), 484 or 0.208 metric is the user-definable variable shape factor, A is the area of the catchment (sq. kilometers – metric; sq. miles), P_X is the excess precipitation-total (1 mm; 1 inch), and T_p is the time to peak (hrs). The time to peak (T_p) and the base time (B_T) determine the characteristics of the unit hydrograph. These values are computed via Equations 3.12 and 3.13, respectively, as follows:

$$T_P = \left[\frac{d + t_c}{1.7}\right],\qquad(3.12)$$

$$B_T = T_P(2.67),\qquad(3.13)$$

where T_p and t_c are the time to peak and time of concentration (hrs), respectively, t_c is 1.67 (L), L is the lag time, d is the time interval (hrs), and B_T is the base time (hrs).

Equation 3.14 describes the relationship between accumulated rainfall (P), initial abstraction (I_a), which includes surface storage, interception, and infiltration before runoff occurs, and potential maximum retention (S). It has been used by the NRCS to compute the direct runoff that occurs as a result of a 24-hour storm rainfall event:

$$Q = \frac{(P - I_a)^2}{(P - I_a) + s},$$ (3.14)

where Q is the accumulated direct runoff (mm).

The correlation between I_a and S was established based on empirical data collected from watershed experiments. It eliminates the need to estimate I_a for typical applications. The mathematical relationship utilized in the NRCS runoff equation (Equation 3.15) is as follows:

$$I_a = (0.2S),$$ (3.15)

Substituting 0.2S for I_a in Equation 3.15, the NRCS rainfall-runoff equation becomes the following (Equations 3.16 and 3.17):

$$Q = \frac{(P - 0.2S)^2}{(P + 0.8S)},$$ (3.16)

$$S = \frac{1000}{CN} - 10,$$ (3.17)

where Q is the runoff depth (in), P is the rainfall depth (in), I_a is the initial abstraction (in), S is the potential maximum retention after runoff begins (in), and CN is the curve number.

The NRCS has created an empirical correlation between lag time (L) and time of concentration (T_C), which may be determined using Equation 3.18:

$$d\text{-}d.L = 0.6T_C,$$ (3.18)

The curve number method can estimate the depth of direct runoff and the lag time of a watershed in drainage basins with a watershed area of less than 2,000 acres, where no stormwater runoff measurements have been taken, provided the index describing runoff response characteristics is available. Equations 3.19 and 3.20 can calculate the lag time of the watershed:

$$L = \frac{(L_m^{0.8})\left[\frac{S}{2.54}+1\right]^{0.7}}{735Y^{0.5}} \dots\dots\dots\dots\dots\dots\dots \text{Metric} \qquad (3.19)$$

$$L = \frac{(L_m^{0.8})[S+1]^{0.7}}{1900Y^{0.5}} \dots\dots\dots\dots\dots \text{Imperial system} \qquad (3.20)$$

where L is the lag time (hrs), L_m is the hydraulic length (m or ft. U.S.), Y is the average watershed slope (%), and CN is the NRCS curve number in $S = 25.4 \left[\frac{1000}{CN} - 10\right]$ (mm) and $S = \left[\frac{1000}{CN}\right] - 10$ (in).

Table 3.7 lists some of the most regularly used hydrology and hydraulic analysis and water quality modeling software and design tools that are widely accepted in the engineering industry.

Table 3.7 Hydrology, hydraulic, and water quality modeling software and design tools

Name of Software and Design Tools	Minor Drainage System	Major Drainage System	Hydrology	Hydraulic	Water Quality	Simulation Type	BMP's Built-in
WinTR-20	X		X			Event	No
TR-55			X			Event	No
HGC-GeoHMS	X	X	X			Event	No
HEC-1	X	X	X			Event	No
HEC-HMS		X	X	X		Event/Continuous	No
EPA SWMM	X	X	X	X	X	Event/Continuous	Yes
SUSTAN					X	Event/Continuous	Yes
PondPACK	X		X	X		Event	Yes
Hydro CAD	X		X			Event	Yes
HEC-2	X	X		X			No
HEC-RAS		X		X		Event/Continuous	No
InfoWorks ICM		X	X	X	X	Event/Continuous	Yes
XPSTORM	X	X	X	X		Event	No
WinHSPF	X		X	X	X	Event/Continuous	Yes
InfoSWMM	X	X	X	X	X	Event/Continuous	Yes
MIKE URBAN	X	X	X	X	X	Event/Continuous	Yes
P8	X	X			X	Event/Continuous	Yes
CivilStorm	X	X	X	X		Event	Yes
FlowMaster	X	X	X	X		Event	No
PCSWMM	X	X	X	X	X		Yes
CulvertMaster	X			X		Event	No
FHWA HY-8	X			X		Event/Continuous	No
QUAL2E/QUAL2K		X		X	X	Event/Continuous	No

3.6 CHAPTER SUMMARY

Assessing drainage design necessitates determining the maximum flow rate and volume of stormwater runoff generated within a watershed during a specific storm event. The dimensions of precipitation events, flow conditions, and runoff flow rate significantly influence the dimensions, layout, and functionality of storm drainage and flood control systems.

Urban drainage infrastructures fall into two main categories: minor and major drainage systems. These categories are defined by specific design criteria influenced by factors such as public health, safety, welfare, and economic considerations. The minor drainage system is designed to manage moderate flooding events effectively. Temporary inundation of local streets due to infrequent, intense precipitation events may be deemed acceptable for minor drainage systems over short durations, provided public well-being, safety, and structural protection are maintained. The major system is typically engineered to a higher standard to minimize the impact of infrequent but more intense flood events due to the increased potential risk to public health, safety, and welfare along significant watercourses.

Detention/retention basins, along with other structural and non-structural stormwater practices, are engineered to mitigate the potential increase in peak flow resulting from storms of varying magnitudes, including 1-, 2-, 5-, 10-, 25-, 50-, and 100-year events. Detention structures are typically located at the lowest elevations of the development site and are designed to release stormwater runoff into a public right-of-way or drainage easement. Stormwater treatment practices include structural controls primarily engineered to reduce pollutants and minimize downstream water quality impacts resulting from stormwater runoff. These practices can also provide additional benefits, such as groundwater replenishment, peak runoff attenuation, and stream channel protection.

3.7 CHAPTER PROBLEMS

1. Define the drainage system. What is the difference between major and minor urban drainage systems?

2. The catchment area of the forthcoming development project, "New Normal 2024," experiences a preconstruction stormwater runoff rate of 1.5 cubic meters per second (53 cubic feet per second) during a 10-year storm event. The projected post-construction 10-year stormwater discharge rate is 2.5 cubic meters per second (85 cubic feet per second). The predicted amount of runoff volume is 4.0 inches for this specific area with a Type III rainfall distribution (see Chapter 2). Calculate the required storage capacity needed to restore the flow to its preconstruction conditions.

3. Underground drainage convenience systems (UDCSs) are utilized in minor stormwater drainage systems to facilitate the conveyance of stormwater runoff from roads and other inlets to the major drainage system. These systems play a pivotal role in managing stormwater effectively. Outline and provide an explanation of the various UDCS components.

4. What distinguishes structural from nonstructural stormwater control BMPs? Provide an example.

5. List the stakeholders to consider when managing stormwater runoff and pollution. What are their interests?

6. Determine the overland flow time and runoff discharge rate for a land development project, given the flow path length (L) of 55 meters, slope (S) of 0.04, Manning's roughness coefficient (n) of 0.20, and runoff coefficient (C) of 0.45. The design frequency is set at 100 years.

7. Determine the peak runoff discharge rate and travel time for a roadside gutter for a street improvement project with a flow path length (L) of 100 meters, longitudinal slope (S) of 0.025, Manning's roughness coefficient (n) of 0.013, pavement cross-slope (Sx) of 0.02, and depth at the curb face (d) of 16 centimeters with the water at the curb for the spread (T) of 5 meters. The design frequency is set at 10 years.

8. The "New Normal 2024" residential project is a single-family development covering an area of 0.85 hectares. It includes the construction of a new asphalt street with a concrete curb and gutter. The flow path length (L) of the street is 100 meters and it has a longitudinal slope (S) of 1.5%. The Manning's roughness coefficient (n) for the street is 0.013. Additionally, the pavement has a cross slope (Sx) of 0.02. The depth at the curb face (d) is 0.5 feet and the water spread (T) at the curb is 5 meters.

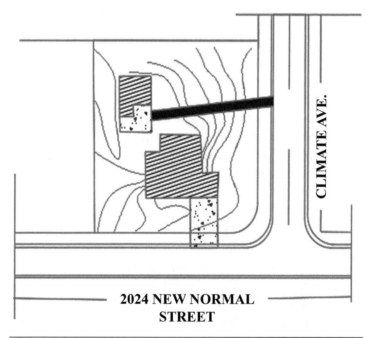

Figure 3.20   "New Normal" street

This project also includes an asphalt parking lot, measuring 100 meters in length and 50 meters in width. The parking lot has a longitudinal slope of 1.5% along its length and a transverse slope of 1%. A continuous gutter is constructed along its length, which then directs water flow into a curb-opening inlet located at the end. Calculate the peak runoff and storage volume using the rational, modified rational, and NRCS hydrograph methods. Given the specified return periods of 2, 5, and 10 years, the rainfall intensity values of 45, 75, and 100 millimeters per hour are associated with a seven-minute storm event. The preconstruction stormwater runoff rate of 2.5 cubic meters per second and the percentage of land covered by the project are as follows:

Asphalt street	3.5%
Concrete curb and gutter	0.25%
Parking lot	5.0%
Single-family house	30%
Driveway	2.5%
Landscape area	58.75%

9. Table 3.8 provides a 15-minute unit hydrograph for the 3-square-kilometer "New Normal" catchment area. Calculate the runoff hydrograph for a rainfall excess of 4.5 cm and the runoff hydrograph for a 20-minute rainfall excess of 6.5 cm.

Table 3.8 Fifteen-minute unit hydrograph for the "New Normal" catchment area

Time (min)	0	15	60	75	90	105	120	130
Discharge (m³/sec)	0	5	8.5	4.5	2.5	1.0	0.5	0

SELECTED SOURCES AND REFERENCES

Albano, R., L. Mancusi, and A. Abbate. 2017. "Improving Flood Risk Analysis for Effectively Supporting the Implementation of Flood Risk Management Plans: The Case Study of 'Serio' Valley." *Environmental Science & Policy* 75: 158–172.

Alfieri, L., P. Burek, L. Feyen, and G. Forzieri. 2015. "Global Warming Increases the Frequency of River Floods in Europe." *Hydrology and Earth System Sciences* 19: 2247–2260.

Alfieri, L., P. Salamon, A. Bianchi, J. Neal, P. Bates, and L. Feyen. 2014. "Advances in Pan-European Flood Hazard Mapping." *Hydrological Processes* 28: 4067–4077.

APWA (American Public Works Association). 1976. History of Public Works in the United States: 1776–1976. Chicago, IL.

Araya, F. and S. Vasquez. 2022. "Challenges, Drivers, and Benefits to Integrated Infrastructure Management of Water, Wastewater, Stormwater and Transportation Systems." *Sustainable Cities and Society* 82: 103913.

Barbosa, A. E., J. N. Fernandes, and L. M. David. 2012. "Key Issues for Sustainable Urban Stormwater Management." *Water Research* 46, no. 20: 6787–6798.

Bárdossy, A. 2007. "Calibration of Hydrological Model Parameters for Ungauged Catchments." *Hydrology and Earth System Sciences Discussions*, 11: 703–710.

Bishop, M. 1968. *The Middle Ages*. Houghton Mifflin Company, Boston.

Brown, S. A., S. Stein, and J. C. Warner. 2001. "Urban Drainage Design Manual: Hydraulic Engineering Circular 22." No. FHWA-NHI-01-021. United States. Federal Highway Administration. Office of Bridge Technology.

Buerger, C. 1915. "A Method of Determining Stormwater Run-Off." Transactions of the American Society of Civil Engineers, 78: 1139–1205.

Burian, S. J. 2001. "Developments in Water Supply and Wastewater Management in the United States During the 19th Century." *Water Resources Impact*, 3(5): 14–18.

Burian, S. J. and F. G. Edwards. 2002. "Historical Perspectives of Urban Drainage." In *Global Solutions for Urban Drainage*, 1–16.

Burian, S. J., S. J. Nix, S. R. Durrans, R. E. Pitt, C.-Y. Fan, and R. Field. 1999. "Historical Development of Wet-Weather Flow Management." *Journal of Water Resources Planning and Management*, 125(1): 3–11.

Butler, D. and J. Parkinson. 1997. "Towards Sustainable Urban Drainage." *Water Science and Technology* 35, no. 9: 53–63.

Butler, D. and J. W. Davies. 2000. *Urban Drainage*. E & FN Spon, London.

Butler, D., C. Digman, C. Makropoulos, and J. W. Davies. 2018. *Urban Drainage*. CRC Press.

Cain, L. P. 1972. "Raising and Watering a City: Ellis Sylvester Chesbrough and Chicago's First Sanitation System." *Technology and Culture* 13 (July): 353–372.

Cettner, A., R. Ashley, M. Viklander, and K. Nilsson. 2013. "Stormwater Management and Urban Planning: Lessons from 40 Years of Innovation." *Journal of Environmental Planning and Management* 56, no. 6: 786–801.

Chaplin, S.E. 1999.Cities, Sewers, and Poverty: India's Politics of Sanitation. *Environ. Urban.* 11: 145–158. 131.

Chow, V. T., D. R. Maidment, and L. W. Mays. 1988. *Applied Hydrology*. McGraw-Hill, Inc., pp. 140–147.

Daedalus Informatics Ltd. 2002. "Knossos: The Palace of King Minos." Daedalus web site, updated 1997. Daedalus Informatics Ltd., Athens, Greece.

Dinicola, R. S. 2001. *Validation of a Numerical Modeling Method for Simulating Rainfall-Runoff Relations for Headwater Basins in Western King and Snohomish Counties, Washington*. Water Supply Paper 2495.

Elliott, A. H. and S. A. Trowsdale. 2007. "A Review of Models for Low Impact Urban Stormwater Drainage." *Environmental Modelling & Software* 22, no. 3: 394–405.

Ellis, J. B. and J. Marsalek. 1996. "Overview of Urban Drainage: Environmental Impacts and Concerns, Means of Mitigation and Implementation Policies." *Journal of Hydraulic Research* 34, no. 6: 723–732.

FAA. Federal Aviation Administration. 1970. Circular on airport drainage (Report A/C 050-5320-5B). Washington, D.C.: U.S. Department of Transportation.

Federal Highway Administration (FHWA). 2005a. *Design of Roadside Channels with Flexible Linings*. Hydraulic Engineering Circular No. 15 (HEC-15). Third Edition. Washington, D.C.

FHWA. 2005b. *Hydraulic Design of Highway Culverts*. HDS-5. Washington, D.C.

FHWA. S. L. Douglass and B. M. Webb. 2020. "Highways in the Coastal Environment, Third Edition, Hydraulic Engineering Circular No. 25 (HEC25)." Federal Highway Administration. Accessed December 2021.

Fread, D. L. 1993. "Flow Routing." In *Handbook of Hydrology*. Edited by D. R. Maidment. McGraw-Hill, Inc.

Gayman, M. 1997. A Glimpse into London's Early Sewers. Reprinted from Cleaner Magazine.

Gest, A. P. 1963. Engineering. In: *Our Debt to Greece and Rome*, edited by G. D. Hadzsits and D. M. Robinson, Cooper Square Publishers, Inc., New York.

Goyen, A., B. C. Phillips, and S. Pathiraja. 2014. *Project 13 Stage 3 Report, Urban Rational Method Review*. Australian Rainfall & Runoff (AR&R) Revision Projects. Engineers Australia, Canberra, Australia.

Gray, H. F. 1940. Sewerage in Ancient and Medieval Times. *Sewage Works Journal*, 12: 939–946.

Gray, S. M. Proposed Plan for a Sewerage System and for the Disposal of the Sewage of the City of Providence, 1st ed., Providence Press Company, Providence, RI, USA, 1884, 130.

Guo, J. C. 2003. *Urban Storm Water Design*. Water Resources Publication.

Gupta, K. 2006. Wastewater Disposal in the Major Cities of India. *Int. J. Environ. Pollut.* 28: 57–66. 136.

Hill, D. 1984. *A History of Engineering in Classical and Medieval Times*. Croom Helm Ltd., London.

Hodge, A. T. 1992. *Roman Aqueducts and Water Supply*. Gerald Duckworth & Co. Ltd., London.

Hughes, M. 2013. "The Victorian London Sanitation Projects and the Sanitation of Projects." *Int. J. Proj. Manag.* 31: 682–691. 126.

Hugo, V. 1863. "Les Misérables . . ." C. Lassalle.

Hydrology and Hydraulics Concepts G282.4 Student Manual. June 2009.

Johnson, F. and F. Chang. 1984. Drainage of Highway Pavements HEC No. 12. FHWA-TS-84-202.

Kirby, R. S. and P. G. Laurson. 1932. *The Early Years of Modern Civil Engineering*. Yale.

Kirby, R. S., S. Withington, A. B. Darling, and F. G. Kilgour. 1956. Engineering in History. McGraw-Hill Book Company, Inc., New York, NY.

Krupa, F. 1991. "Paris: Urban Sanitation Before the 20th Century." A History of Invisible Infrastructure.

Le Jallé, C., D. Désille, and G. Burkhardt. 2013. "Urban Stormwater Management in Developing Countries." In *Novatech 2013-8ème Conférence Internationale sur les Techniques et Stratégies Durables pour la Gestion des Eaux Urbaines par Temps de Pluie / 8th International Conference on Planning and Technologies for Sustainable Management of Water in the City*. GRAIE, Lyon, France.

Li, J., D. Joksimovic, and J. Tran. 2015. "A Right-of-Way Stormwater Low Impact Development Practice." *Journal of Water Management Modeling*. doi: 10.14796/JWMM.C390.

Lloyd-Davies, D. E. 1906. The Elimination of Storm Water from Sewerage Systems. Minutes of Proceedings, Institution of Civil Engineers (London), 164: 41–67.

Maner, A. W. 1966. Public Works in Ancient Mesopotamia, Civil Engineering, 36(7): 50–51.

Marsalek, J. 2001. "Review of Stormwater Source Controls in Urban Drainage." *Advances in Urban Stormwater and Agricultural Runoff Source Controls*, 1–15.

Marsalek, J., T. O. Barnwell, W. Geiger, M. Grottker, W. C. Huber, A. J. Saul, and H. C. Torno. 1993. "Urban Drainage Systems: Design and Operation." *Water Science and Technology* 27, no. 12: 31–70.

McMath, R. E. 1887. Determination of the Size of Sewers, Transactions of the American Society of Engineers.

Metcalf, L. and Eddy, H. P. 1928. American Sewerage Practice, Volume I: Design of Sewers. McGraw-Hill Book Company, Inc., New York, NY, pp. 1–33.

Mguni, P., L. Herslund, and M. B. Jensen. 2016. "Sustainable Urban Drainage Systems: Examining the Potential for Green Infrastructure-Based Stormwater Management for Sub-Saharan Cities." *Natural Hazards* 82, no. 2: 241–57.

Mumford, L. 1937. Technics and Civilization. *The Journal of Nervous and Mental Disease*, 86(1): 111–112.

Mumford, L. 1961. *The City in History: Its Origins, Its Transformations, and Its Prospects*. Harcourt, Brace & World, New York, NY.

Niemczynowicz, J. 1997. "State-of-the-Art in Urban Stormwater Design and Research." An invited paper was presented at the Workshop and Inaugural Meetings of the UNESCO Center for Humid Tropics Hydrology, Kuala Lumpur, Malaysia, November 12–14, 1997.

Pazwash, H. 2011. *Urban Storm Water Management*. CRC Press.

Reid, D. 1991. Paris Sewers and Sewer Men. Harvard University Press, Cambridge, MA.

Reynolds, R. 1946. Cleanliness and Godliness. Doubleday and Company, Inc., New York, NY.

Rossman, L. 2017. *Storm Water Management Model Reference Manual Volume II—Hydraulics*. U.S. Environmental Protection Agency, Washington, DC, EPA/600/R-17/111. Accessed at: https://nepis.epa.gov/Exe/ZyPDF.cgi?Dockey=P100S9AS.pdf.

Schladweiler, J. C. 2002. Tracking Down the Roots of Our Sanitary Sewers. In Pipeline 2002: Beneath Our Feet: Challenge and Solutions. Proceedings of the ASCE Pipeline Division of ASCE, Cleveland, OH, USA, 4–7, August 2002; Kurz, G.E., Ed. American Society of Civil Engineers: Reston, VA, USA, pp. 1–27.

Seeger, H. 1999. The History of German Wastewater Treatment. *Eur. Water Manag.* 2, 51–56. 133. Society of Civil Engineers, 16: 179–190.

Shafique, M. 2016. "A Review of the Bioretention System for Sustainable Storm Water Management in Urban Areas." *Materials and Geoenvironment* 63: 227–235.

Sidwick, J. M. 1977. A Brief History of Sewage Treatment. Thunderbird Enterprises, Ltd.

Strong, D. 1968. The Early Etruscans. G. P. Putnam's Sons, New York, NY.

Tarr, J. A. 1979. The Separate vs. Combined Sewer Problem: A Case Study in Urban Technology Design Choice. *Journal of Urban History*, 5(3): 308–339.

Thakali, R., A. Kalra, S. Ahmad, and K. Qaiser. 2018. "Management of an Urban Stormwater System Using Projected Future Scenarios of Climate Models: A Watershed-Based Modeling Approach." *Open Water* 5, no. 2: 1.

USA Cast Iron Pipe & Foundry Co. 1914. Cast Iron Pipe, Standard Specifications Dimensions and Weights; USA Cast Iron Pipe & Foundry Co., Burlington, VT.

U.S.D.A.-Natural Resource Conservation Service (NRCS). 1956. *National Engineering Handbook, Section 5: Hydraulics*. Washington, D.C.

U.S.D.A.-Natural Resource Conservation Service (NRCS). 1982. *Project Formulation—Hydrology*. Technical Release No. 20 (TR-20). Washington, D.C.

U.S.D.A.-Natural Resource Conservation Service (NRCS). 1985. *National Engineering Handbook, Part 630: Hydrology*. Washington, D.C.

U.S.D.A.-Natural Resource Conservation Service (NRCS). 1986. *Urban Hydrology for Small Watersheds*. Technical Release No. 55 (TR-55). Washington, D.C.

Uyumaz, A. 1994. "Highway Storm Drainage with Kerb-Opening Inlets." *Science of the Total Environment* 146: 471–478.

Walesh, S. G. 1991. *Urban Surface Water Management*. John Wiley & Sons.

Walker, D. 1987. Great Engineers: The Art of British Engineers, 1837–1987. St. Martin's Press, New York, NY.

Waring, G. E. 1875. "The Sanitary Drainage of Houses and Towns." *Atlantic Monthly* 36 (November 1875): 535–553.

Webster, C. 1962. The Sewers of Mohenjo-Daro. *Journal Water Pollution Control Federation*, 34(2): 116–123.

Winter, K., and S. Mgese. 2011. "Stormwater Drainage: A Convenient Conduit for the Discharge of Urban Effluent into the Berg River, South Africa." In *12th International Conference on Urban Drainage, Porto Alegre, Brazil*.

Wright, K. R. and A. Valencia Zegarra. 2000. Machu Picchu Is a Civil Engineering Marvel. ASCE Press, American Society of Civil Engineers, Reston, VA.

Zeng, Z., X. Yuan, J. Liang, and Y. Li. 2021. "Designing and Implementing an SWMM-Based Web Service Framework to Provide Decision Support for Real-Time Urban Stormwater Management." *Environmental Modelling & Software* 135: 104887.

4

RESETTING URBAN DRAINAGE SYSTEM DESIGN CRITERIA FOR THE "NEW NORMAL"

4.1 INTRODUCTION

Recent catastrophic events have left indelible marks on various regions worldwide. After enduring recurring floods, states in the U.S., such as California, Louisiana, Texas, and Maryland, along with countries like Pakistan, Bangladesh, and India, have experienced the severe impacts of these extreme weather events. Large-scale levee breaches and overtopping across the Midwestern U.S., including Nebraska, Missouri, South Dakota, Iowa, and Kansas, have posed significant threats to economies, public health, and natural ecosystems.

Understanding the gravity of these "New Normal" weather patterns is crucial, particularly the implications for our cities and infrastructure systems. Design storm criteria, referring to specific intensity and/or frequency, play a pivotal role in overcoming the challenges faced by infrastructure systems. The current design standards for urban drainage systems will not be suitable in an era of growing complexity, unpredictability, and extreme weather events due to climate change.

The efficacy of different drainage infrastructure design methodologies and their underlying assumptions has faced growing scrutiny, especially due to the occasional use of inaccurate or inconsistent historical data. Previous urban drainage system designs were based on the assumption of a consistent climate, where historical patterns could reliably predict future conditions. However, the escalating unpredictability and heightened intensity of extreme events challenge the assumption of stationary design storm requirements.

There are many concerns surrounding urban drainage system design due to evolving requirements and criteria. However, there is currently no established solution to effectively tackle the design challenges arising from extreme events. This chapter will examine the fundamental principles underlying the design of current storm drainage infrastructure, address the various challenges associated with the existing design and standards for urban drainage infrastructure, and discuss potential alternatives and enhancements to the current approach. Factors to consider include climate change, the increasing complexity of urban settings, and the interdependence of infrastructure and social systems.

Urban drainage system design criteria comprise specific standards and guidelines that establish the maximum allowable level or probability of hazards that a particular drainage infrastructure may encounter during various events. These criteria are used to design and construct drainage infrastructures to ensure their resilience. Examples of drainage infrastructure systems include large levees, culverts, flood control structures, road drainage systems, airport drainage systems, power supply infrastructure, and buildings.

4.1.1 Interaction Between Minor and Major Drainage Systems

An obvious distinction exists between *minor* and *major* drainage infrastructure systems, and the optimal approach to development involves incorporating these two storm drainage systems.

Minor drainage systems, conventional overland, and underground drainage convenience systems are typically engineered to convey stormwater runoff from less intense and more frequent storm events, such as those occurring every 5 to 10 years. These systems collect stormwater runoff from roadways, parking facilities, and other paved surfaces, efficiently transporting the runoff into inlets, catch basins, manholes, and pipes, leading to the major drainage infrastructure systems. However, due to financial constraints, these systems have a limited hydraulic capacity to manage extreme weather or "New Normal" peak discharges effectively.

Major drainage infrastructure systems transport stormwater runoff from infrequent and severe weather events when the minor storm drainage system's capacity is exceeded. These systems typically consist of large pipes, drainage channels, stormwater storage areas, watercourses, thoroughfares, and basins designed for temporary water storage. The implementation of a *split* design philosophy in drainage infrastructure requires consideration and mitigation of potential consequences during the design phase, such as surcharging of the minor drainage system leading to property flooding.

In urbanized areas, installing major drainage infrastructure and improving the hydraulic capacity of its key components can present significant challenges. These challenges include increasing pipe capacity, intercepting peak stormwater runoff discharge, managing volume through retention and detention, creating an extended storage zone for stormwater runoff volume, and retrofitting. However, these processes can be efficiently completed during the preliminary planning and design phases of new development projects. Therefore, it is critical to conduct a thorough assessment of stormwater peak runoff and volume management requirements, as well as potential enhancements to the capacity of existing drainage infrastructure to handle extreme weather events. The evaluation process should include an assessment of space distribution along both current and projected drainage routes, as well as flood-prone zones at lower elevations. This analysis aims to mitigate and minimize any potential damage to private property or critical public infrastructure by strategically positioning various elements, such as curb jumps, surface drainage channels, and green street median bioretention zones, to manage and mitigate flooding levels on streets and roads effectively, as well as intercept peak stormwater runoff and volume.

4.2 OVERVIEW OF CURRENT URBAN DRAINAGE SYSTEM DESIGNS AND PLANNING APPROACHES

Numerous physical factors significantly influence the design process of urban drainage infrastructure. These processes typically involve determining the optimal sizes for system components necessary to manage a designated design event's peak stormwater runoff discharge or volume effectively. The value of this design peak discharge or volume depends on the intended level of service for a specific drainage infrastructure system. This task requires assessing the potential risk of flooding and determining the acceptable frequency of such events. The design flow magnitude is usually expressed in terms of the likelihood of flow exceeding a specific threshold or the frequency at which similar magnitudes occur. Once the desired level of service and appropriate design frequency have been selected, the technical design process begins. This process includes applying established design methodologies, evaluating hydrologic input data, and identifying suitable design parameters. It also involves allocating land and easements to accommodate the required drainage infrastructure and implementing measures to regulate or restrict the interaction between the pre- and post-development drainage infrastructure and the surrounding development.

There are several well-established methodologies available for calculating peak stormwater runoff discharge or volume. However, the primary methodologies used to predict the peak flow rate or volume of stormwater runoff in catchment areas, where hydrology and hydraulics are studied, are empirical methods and hydrologic simulation models. These systems are based on parameters representing the land use on and off the site of the new development being constructed, as well as rainfall and snowmelt values suitable for the specific local conditions. Numerous factors, such as the imperviousness ratio of the drainage area, the nature of the soil, and the extent of vegetation cover, affect land use. To accommodate the various design parameters arising from site conditions, both methodologies use a range of design parameters instead of single values.

The selection of appropriate parameters, such as precipitation and snowmelt for a particular watershed during hydrologic and hydraulic analysis, is a critical component in planning and designing stormwater drainage infrastructure networks. However, these tasks are challenging due to significant variations in climatic variables, both spatially and temporally, due to climate change. Various methodologies are frequently used, including intensity-duration-frequency (IDF) curves, historical design storms, and synthetic design storms. IDF curves are a widely used technique to represent the statistical distribution of extreme precipitation data collected at a specific geographical point. Historic design storms are typically characterized as the most extreme storm events recorded at a specific geographic site.

The generation of synthetic design storms involves calculating average precipitation patterns using historical data from a larger watershed. However, decision makers must recognize that the methods for selecting precipitation and snowmelt design data rely on historical stationary data. The assumptions used to forecast future IDF curves, based on the historical intensity data of design rainfall and snowmelt over a period, is no longer valid due to the nonstationary data caused by climate change. Modifying or removing these assumptions could significantly improve the design and planning of stormwater drainage infrastructure. This is because the sizing, upgrading, and retrofitting of drainage infrastructures will be designed with the assumption of both historical stationarity and nonstationary data.

4.2.1 Challenges with Conventional Urban Drainage System Designs and Planning

The conventional approach to designing and planning urban drainage systems does not sufficiently account for the effects of shifting precipitation patterns on drainage infrastructure. Historical IDF curves must be adjusted to protect the public's financial, environmental, and societal interests in the short and long term. Given the projected escalation in the severity and unpredictability of climate hazards, such as extreme precipitation events, substantial challenges and inconsistencies can be expected in the current urban drainage infrastructure design standards.

The nonstationarity characteristics of precipitation data, which are often used in the design of drainage infrastructure, are frequently overlooked in hydrological and hydraulic analyses. This neglect of the nonstationarity principle is a significant oversight. This principle underscores that systems are not static and unchanging; instead, they display variations within a continually changing range of variability. Ignoring this issue is likely to significantly impact future challenges that will emerge or be exacerbated by climate variability and uncertainty. Moreover, it is highly likely that this will pose considerable challenges in the formulation of stormwater management regulations and the design of drainage infrastructure systems. Drainage infrastructure systems that were designed using stationary rainfall data may encounter a situation where they have been engineered to handle conditions that are no longer relevant. Consequently, historical data may no longer serve as a reliable predictor of future outcomes.

Dynamic conditions resulting from climate change can compromise the reliability of the data that underpin a large part of the quantitative understanding and mitigation of potential risks to drainage infrastructure. The increasing frequency and intensity of extreme weather events can diminish the ability of infrastructure services and plans to function effectively. Thus, design criteria that rely heavily on historical return periods and data, such as the rational method, may see a decrease in their validity. Efforts should be focused on developing and implementing design standards that adopt a proactive approach and have the ability to efficiently manage events associated with nonstationarity. In addition to the challenges posed by climate nonstationarity, the increasing complexity of urban and infrastructure systems presents a significant hurdle for existing design and implementation frameworks.

Urban and infrastructure systems display complex interdependencies among various elements, users, and administrators. These systems are intricately linked with other community, environmental, and technological frameworks. The process of integrating and connecting systems, both internally and externally, often leads to increased complexity, which complicates the identification and manipulation of individual causal relationships among system components. Thus, the systems exhibit emergent properties in terms of their behavior and performance. Furthermore, the dynamic nature of urban infrastructure systems presents significant challenges in effectively implementing risk analysis and management strategies. As specific cause-and-effect relationships become less clear, the likelihood and potential consequences of a disruption become more difficult to determine and anticipate.

4.3 PLANNERS, DESIGNERS, AND STAKEHOLDERS IN EXTREME WEATHER-BASED DESIGN STANDARDS

Implementing design standards that consider extreme weather conditions is a complex task for planners, designers, and stakeholders. This task requires these diverse stakeholders to navigate complex decision-making processes and collaborate effectively. A primary strategy for continually developing stormwater infrastructure system criteria, given the increasing challenges, is to adopt an interdisciplinary, multi-hazard, and inclusive approach that considers multiple perspectives.

A well-structured and transparent process is vital for effective stakeholder engagement. Numerous organizations are currently involved in implementing robust drainage infrastructure standards, governed by legal frameworks, formal agreements, and informal arrangements. The process is complex and is further complicated by above-ground flows.

4.3.1 Stakeholder Engagement

Effective stakeholder engagement is critical for the successful implementation of extreme weather-based design standards and policies. Stakeholder and the public's involvement is crucial in ensuring the optimal functioning of such standards and policies. Engaging stakeholders from both the private and nonprofit sectors allows policymakers to harness a wide range of expertise and secure support from various political entities. Public participation enhances public awareness of climate change risks, encourages political support for selected policies addressing this issue, and provides legitimacy to the funds allocated for adaptation measures. Table 4.1 presents a comprehensive summary of the diverse stakeholders involved and includes representative examples.

Table 4.1 Types of stakeholders

Internal	Examples
Project resources	Project owners, financiers/creditors, employees, client organizations (government, state), shareholders, project managers, board members
External	**Examples**
Public sector	Authorities (city, state, federal), governments, ministries concerned (e.g., department of transportation, covering air, surface, and sea), electricity authority, urban development, public works, environmental agencies, labor unions
Suppliers/market players	Inland and sea transport operators, logistics companies, shipping and transportation companies
Community interest groups	Social organizations, consumers and taxpayers, nonprofit organizations, environmental groups, the media

4.3.2 Collaboration with Government Entities

Government entities have unique capabilities to support communities in developing strategies to adapt effectively to climate change impacts, beyond simply funding these types of projects. The government provides essential resources, such as climate and weather data, which are useful for conducting vulnerability and risk assessments. Planners and designers involved in community adaptation must have access to these resources. Moreover, governments play a significant role in facilitating community efforts to strategize and prepare for climate change impacts. It is crucial to utilize available resources to ensure that all communities proactively adapt to climate change. Officials at all government levels should maintain ongoing collaborations to ensure effective coordination of efforts and optimal resource utilization.

4.3.3 Addressing Current Vulnerability and Risk

Professionals involved in the design and planning of new norms must assess the existing vulnerability levels among individuals, considering factors such as sensitivity, adaptability, and the potential impact of current policies and practices. This information forms the basis for a comprehensive risk assessment, which evaluates the likelihood of experiencing extreme weather events and their associated impacts, such as sea-level rise. The primary objective is to assess the community's risk level associated with climate change. After a thorough evaluation of vulnerability and risk, planners can develop new strategies. For a comprehensive assessment of vulnerability and risk, communities must have access to accurate information provided by high-level government entities.

4.3.4 Planning

Planning equips communities with a variety of tools to prepare for severe weather. By adopting an adaptation perspective, communities can make informed decisions that enhance resilience to extreme weather events, strengthen adaptive capacities, and mitigate potential disaster risks. Local land-use plans can effectively prevent or restrict development in hazard-prone areas, such as floodplains, reducing the vulnerability of individuals and assets. Strategic placement of infrastructure and transportation routes can mitigate the vulnerability of key assets to potential damage caused by the "New Normal."

4.3.5 Rules and Regulations

One of the primary ways governments can mitigate losses from extreme weather events is through regulatory instruments, such as codes and standards. An urban drainage infrastructure system, a network of structures and facilities managing stormwater and wastewater flow in urban areas, relies on these rules and regulations for proper design and construction methodologies. When recognizing climate change as a potential risk, codes and standards can serve as valuable tools for efficient adaptation. In many urban areas, the Department of Public Works plays a central role in establishing and revising drainage design standards and guidelines. These updates provide an opportunity to integrate advanced measures into design and construction procedures.

4.3.6 Insurance

The insurance industry has a proven track record of effectively mitigating climate change risks and is well-positioned to support communities in their adaptation efforts. During the reconstruction or rehabilitation of insured infrastructures damaged by extreme weather events, various adjustments can enhance their resilience to potential future hazards. Implementing variable, risk-based insurance premiums can incentivize individuals and organizations to mitigate risk or relocate from high-risk regions. However, neglecting the vulnerability of insured assets could result in persistent and severe losses, undermining the effectiveness of this tool. The availability and extent of insurance coverage for an asset can be limited or unavailable depending on its location, characteristics, and associated risks. Therefore, integrating insurance within climate change risk management strategies is crucial.

The integration of new climate change policies into routine decision making and behaviors enhances their effectiveness. Decisions about land use, building placement, and similar factors carry significant long-term implications. Ignoring the potential impact of future extreme weather events could increase a community's vulnerability. Local governments and municipalities have found that forming an internal steering committee to integrate new information into existing policies and programs, as well as embedding adaptation principles into official community plans, are effective strategies for mainstreaming. Incorporating climate change data into municipal policies and programs is crucial, as it can significantly influence community resilience.

The presence of severe meteorological phenomena poses substantial risks to the welfare and security of the population. Therefore, governments must adopt a deliberate and coordinated strategy to reduce community susceptibility and enhance their capacity to manage these hazards. While most instrumental adaptation measures are implemented at the local level, they should be part of a comprehensive public policy strategy aimed at improving overall community resilience. This requires coordinated efforts across all government levels. The objectives, principles, and tools discussed in this book aim to guide the development of climate policies and propose a framework for intergovernmental support in local climate adaptation efforts.

4.4 ADAPTING DRAINAGE DESIGN GUIDELINES TO CLIMATE CHANGE

Climate change considerations should be integrated into current drainage design guidelines and principles. The ongoing refinement of design criteria for drainage infrastructure systems is crucial to effectively address the escalating threats posed by climate change and the increasing complexity of extreme weather phenomena. These criteria should adopt a multidisciplinary approach, consider various hazards, and promote inclusivity to adequately respond to the evolving needs of drainage systems in the context of the "New Normal" (see Figure 4.1).

Efforts should focus on enhancing the integration of uncertainty, complexity, and adaptability into storm criteria and risk assessment for urban infrastructure design. While return-period and threshold-based approaches are expected to remain relevant for specific components and subsystems, they may require more frequent revisions and updates. Where applicable and relevant, supplementary safety measures should be incorporated, and thresholds and return periods for infrastructure design and construction should be enhanced. This is particularly important when considering the effects of global warming and climate variations. However, system-level strategies also need to be adapted and differentiated to lower the risk of future damage to infrastructure.

Figure 4.1   In October 2022, a 25-year storm event washed out a culvert on Pusha Rd. in Glenburn, ME. This incident led to a 5-mile detour and required $1,111,000 for repairs. Source: Maine Department of Transportation.

It is critical to carefully consider equity and the potential consequences of failure. One potential strategy to improve existing infrastructure-centric design approaches is to integrate nature-based solutions and green infrastructure within urban areas. Risk analysis and reasoning can be used to strike a balance between *safe-to-fail* and *fail-safe* approaches. This involves focusing on preventing disruptions while recognizing that some level of disruption is inevitable. The primary objective is to minimize the impact of these disruptions. Prioritizing the well-being of communities and individuals over infrastructure and potential risks is essential. This includes ensuring access to necessary services and addressing any adverse effects they may encounter. Governments should incorporate experimental capabilities, continuously evaluate performance, and consistently update regulations, guidelines, and methodologies. It is widely recommended that governments move away from relying solely on design criteria based on the concept of temporal stationarity. To develop extensive solution spaces, policymakers need to broaden their understanding of urban and design domains. This expansion should include not only technical aspects but also social, ecological, and technical factors. By broadening their knowledge and understanding, policymakers can effectively address the complex challenges that will arise in the coming decades.

4.4.1 Extreme Rainfall Event Management in Urban Areas

Extreme precipitation events primarily result in potential flooding. This risk may be amplified in urban areas where impermeable pavements facilitate rapid water drainage into sewage systems. Additionally, heavy rainfall increases the likelihood of landslides, which occur when the water table rises, saturating the ground and compromising slope stability. Excessive precipitation can also degrade water quality, adversely affecting human well-being and ecological systems. Stormwater runoff, often containing pollutants such as heavy metals and pesticides, can contaminate freshwater environments, negatively impacting aquatic ecosystems and reducing water quality for human use. Currently, no single approach has been definitively effective in addressing the impacts of intense precipitation on urban environments. However, the integration of Chicago design storms, empirical area reduction factors (ARFs), and intensity-duration-area (IDA) curves can provide a fundamental basis for evaluating the effectiveness of various methods in managing high-intensity rainfall events and adaptation scenarios.

ARFs have been developed for numerous global locations based on empirical data (NERC 1975). Theoretical literature has established a connection between rainfall spatial correlation and ARFs. Rodrigues-Iturbe and Mejia (1974) initiated this investigation, later expanded by Bras and Rodriguez-Iturbe (1993). Asquith and Famiglietti (2000) examined annual peak storm events and investigated ARFs. De Michele et al. (2001) developed universal ARFs. Equation 4.1 can be used to calculate ARFs, which represent the geographical correlation as a function of rainfall duration:

$$ARF = C_0 \, exp\left[-C_1\left[\left(\tfrac{A}{L^2}\right)^{n}\right]\right], \qquad (4.1)$$

where A represents the area, L represents the spatial correlation length that is dependent on the duration of rainfall, and C_0, C_1, and n represent the coefficients used for calibration.

The combination of an IDF curve, representing the rainfall intensity for a given return period, with ARF curves can produce IDA curves. IDA curves are mathematical models that determine the rainfall intensity for a specific return period, considering the duration and area of the rainfall event.

A cumulative duration series (CDS), as described by Keifer and Chu (1957), is characterized by the inclusion of T-year storms of various durations within a single storm event. In practice, these storms can range from one minute to a predetermined maximum duration, typically several hours. The IDA curve for a specific return period T is represented as follows:

$$i\left(t_d; T\right) = \frac{a}{t_d^{b}+c}. \qquad (4.2)$$

CDS curves can be mathematically represented as shown in Equations 4.3 and 4.4 (Keifer and Chu 1957), which correspond to the CDS curve before and after peak flow, respectively:

$$i\left(t_d; T\right) = \frac{a\left((1-b)\left(\frac{t_b}{r}\right)^{b}+c\right)}{\left(\left(\frac{t_b}{r}\right)^{b}+c\right)^{2}}, \qquad (4.3)$$

$$\frac{a\left((1-b)\left(\frac{t_b}{1-r}\right)^{b}+c\right)}{\left(\left(\frac{t_b}{1-r}\right)^{b}+c\right)^{2}}. \qquad (4.4)$$

The parameters a, b, and c in the IDA curve are associated with the return period T. t_d represents the duration of rainfall, while r represents the fraction of t_d that occurs before the peak.

4.4.2 Design Return Intervals in Nonstationary Conditions

The complexity of this cross-disciplinary approach presents a significant challenge in managing the effects of extreme events in established urban areas. Integrated analyses necessitate the establishment of several key assumptions. The analysis of runoff resulting from individual precipitation events is significantly influenced by the precipitation's inherent spatiotemporal attributes. Antecedent conditions may also be significant. Software used for modeling inundations must be capable of accepting rainfall input with high spatiotemporal resolution.

In evaluating the typical characteristics of runoff resulting from extreme precipitation events in urban catchments, the use of basic point estimates of IDF remains a leading-edge methodology. This methodology involves using CDS as the primary input for modeling urban inundation. The CDS is a human-made precipitation event designed to simulate the inundation of a sewage system that aligns with a predetermined recurrence interval for the entire metropolitan drainage area. Regional IDF correlations must be considered to estimate the CDS. These correlations depend on various rainfall parameters, such as mean annual precipitation, rainfall location and duration, and return period. The inclusion of these parameters is crucial for accurately estimating the CDS.

The fundamental premise of using CDS is that the catchment's antecedent conditions have a negligible impact on the computed magnitude of floods resulting from intense precipitation events. When designing a stormwater drainage system, it is important to ensure that the precipitation with a recurrence interval of T years corresponds to the runoff with the same recurrence interval T. Implementing this approach allows designers to effectively strategize and develop drainage systems that can accommodate the expected peak stormwater runoff or volume during such events. This assumption implies a linear relationship between rainfall and runoff.

4.4.3 Stationary and Nonstationary Flood-Frequency Analysis

In flood-frequency analysis, it is essential to assume that annual maximum flood discharges are independent and identically distributed random variables. This assumption implies that the statistical properties of flood data remain consistent over time. Consequently, established statistical methods can be employed to calculate the annual probability of a flood of a specific magnitude or to determine the magnitude of a flood with a particular recurrence interval. The statistical representations of flow characteristics and flood-frequency analysis play a crucial role in urban drainage system planning and design. Therefore, it is vital to consider methods that include both stationary and nonstationary hydrologic processes, especially in light of extreme weather events or hydrological extremes associated with climate change.

Hydrologic extremes are becoming increasingly frequent, with no discernible trend in their occurrence. This unpredictability may lead to an underestimation of future extremes, posing a significant challenge in designing adequate drainage systems and accurately assessing potential risks. This uncertainty can also complicate decision-making processes in managing hydrologic extremes. There is a growing need to adapt traditional drainage system design approaches to accommodate nonstationary flood-frequency analysis, which involves understanding and incorporating the dynamic nature of hydrologic processes.

Nonstationarity refers to the concept that the statistical properties of hydrological variables, such as rainfall and streamflow, can change over time. This contrasts with the traditional assumption of stationarity, which posits that these variables follow a fixed pattern. By integrating nonstationary flood-frequency analysis into drainage system design, engineers and planners can better account for the changing nature of hydrologic processes. This integration allows for a more accurate evaluation of flood hazards and the development of efficient drainage systems to mitigate these hazards. Adapting traditional drainage system design approaches to include nonstationary flood-frequency analysis requires a shift in perspective and methodology. It necessitates considering historical data, climate change projections, and other factors influencing the changing patterns of rainfall and streamflow, which are essential for effectively understanding and managing hydrologic processes.

It is important to note that *stationarity* refers to the idea that natural systems fluctuate within a consistent range of variability without significant changes over time. For instance, a hydrologic time series can be considered stationary if it does not exhibit any trends, shifts, or periodic patterns. Conversely, *nonstationarity* refers to the phenomenon where the statistical characteristics of natural systems change over time.

4.4.4 Generalized Extreme Value (GEV) Model for Stationary and Nonstationary Flood-Frequency Analysis

The GEV model is widely used in statistical modeling to analyze and predict extreme weather events. It is particularly effective in identifying the most suitable distribution to represent extreme rainfall patterns across different geographic regions. The GEV model integrates three separate distribution models derived from extreme value theory: the Gumbel, Fréchet, and Weibull distributions. Numerous studies have found that the GEV distribution outperforms other distributions, such as the generalized normal distribution and the Pearson Type III distribution, in describing and predicting annual extreme rainfall. This is attributed to the GEV model distribution's basis in statistical principles related to extreme random variables. However, all three distributions have similar performance in accurately describing the annual maximum rainfall.

4.4.5.1 Stationary Generalized Extreme Value Model

Equation 4.5 provides a mathematical representation of the cumulative probability distribution function for stationary GEV distribution as:

$$f(x) = \begin{cases} exp\left[-\left(1 - \frac{\xi}{\alpha}(x - \mu)^{\frac{1}{\xi}}\right)\right] \; ; \; \xi \neq 0 \\ exp\left[- exp\left(-\left(\frac{x-\mu}{\alpha}\right)\right)\right] \; ; \; \xi = 0 \end{cases}, \quad (4.5)$$

where μ represents the location parameter, α represents the scale parameter, and ξ represents the form parameter. For ξ values less than zero, the GEV distribution is equivalent to the Weibull distribution and can be used to model heavy-tailed behavior. The scenario where $\xi > 0$ corresponds to the Fréchet distribution—employed to model datasets that exhibit a left-skewed pattern. The scenario where $\xi = 0$ corresponds to the Gumbel distribution—characterized by scale and location parameters and has a moderate right tail. The GEV parameters are calculated via the maximum likelihood estimation approach (Equations 4.6 and 4.7):

$$L(\mu, \sigma, \xi) = -n \log \sigma - \left(1 + \frac{1}{\xi}\right) \sum_{i=1}^{n} log \left[1 + \xi \left(\frac{x_i - \mu}{\sigma}\right)\right]$$
$$- \sum_{i=1}^{n} \left[1 + \xi \left(\frac{x_i - \mu}{\sigma}\right)\right]^{-\frac{1}{\xi}} \; ; \; \xi \neq 0 , \quad (4.6)$$

$$L(\mu, \sigma, \xi) = -n \log \sigma - \sum_{i=1}^{n} log \frac{x_i - \mu}{\sigma}$$
$$- \sum_{i=1}^{n} \exp \left\{-\frac{x_i - \mu}{\sigma}\right\} ; \; \xi = 0 . \quad (4.7)$$

4.4.5.2 Nonstationary Generalized Extreme Value Model

The nonstationary GEV model is an advanced version of the stationary GEV model. In this model, the parameters of interest, specifically location (μ) and scale (α), may exhibit temporal variability. The nonstationarity that is observed in a time series can often be attributed to the presence of periodic patterns or a noticeable trend. The location and scale parameters can be mathematically represented as a periodic function with respect to time (t), as shown by Equations 4.8 and 4.9, respectively. The covariates comprise a diverse range of factors that contribute to a comprehensive analysis:

$$\mu(t) = \beta_0 + \frac{\beta_1 sin2t\pi}{6} + \frac{\beta_2 cos2t\pi}{6}, \tag{4.8}$$

$$\mu(t) = \phi_0 + \frac{\phi_1 sin2t\pi}{6} + \frac{\phi_2 cos2t\pi}{6}. \tag{4.9}$$

4.4.5.3 Goodness-of-Fit Tests

Goodness-of-fit tests are statistical tools used to evaluate the alignment of a data set with a specific probability distribution. They assess the compatibility between the data and the distribution, quantifying the agreement or discrepancy between the observed and the expected data. These tests provide insights into the suitability of a chosen probability model for a data set. To analyze a GEV model effectively, it is crucial to ascertain whether the model is stationary or nonstationary. Several tests, including the Akaike Information Criterion (AIC), the Bayesian Information Criterion (BIC), and the Likelihood Ratio (H), are commonly used for this purpose. The Likelihood Ratio test, which compares the likelihood of the observed data under two competing hypotheses, is particularly prevalent. Equations 4.10 and 4.11 represent the fundamental concepts of hypothesis testing and model selection, widely used in statistical analysis:

$$H_0 = Model\ is\ stationary, \tag{4.10}$$

$$H_1 = Model\ is\ nonstationary. \tag{4.11}$$

Equations 4.12 and 4.13 provide the mathematical expressions for the AIC and BIC, respectively. These criteria are used extensively across various disciplines to compare different models based on their goodness of fit and complexity:

$$AIC = -2l + 2p, \qquad (4.12)$$

$$BIC = -2l + 2p \log n. \qquad (4.13)$$

The maximal log-likelihood for this model is denoted as l, the model's parameters as p, and the total number of samples as n. The most suitable model is determined by comparing the stationary and nonstationary GEV models and selecting the one with the smallest value for each criterion. The deviance test statistic, represented as D, is a critical measure used in statistical analysis to evaluate a model's goodness of fit and to compare multiple entities. This comparison process can reveal similarities and differences between these entities, providing valuable insights for decision making.

4.4.5 Retrofitting Existing Drainage Systems

Implementing stormwater retrofits on existing developed sites or as part of redevelopment initiatives is crucial for mitigating the adverse effects of extreme weather events. These effects, specifically, are increased stormwater peak runoff or volume. The anticipated increase in the frequency and intensity of stormwater peak runoff presents a significant challenge for designing stormwater drainage infrastructure. Given the future of extreme weather events, it has become imperative to enhance the capacity of drainage infrastructure in various urban municipalities. However, there are increasing concerns about the long-term sustainability of expanding underground pipeline infrastructure as a climate adaptation strategy, instead of exploring alternative solutions. One such solution is the integration of a decentralized drainage system into the existing storm sewer system. In recent years, the potential benefits of decentralized drainage infrastructure systems have gained recognition. Depending on design choices, decentralized solutions can positively impact sustainable development by enhancing aesthetic value, public perception, and the urban environment. A decentralized drainage system, which deviates from the traditional centralized approach, handles excess stormwater runoff or volume by distributing the responsibility across multiple smaller units.

This approach aims to address the challenges posed by conventional centralized drainage systems. By implementing this approach, we can efficiently and concurrently meet the requirements for climate change adaptation and urban recreational services. This strategy allows us to address two important issues simultaneously: (1) ensuring our cities are resilient to climate change impacts and (2) providing our communities with access to recreational spaces and activities.

Implementing stormwater retrofits often involves facing challenges and obstacles that arise from various site constraints. These constraints can hinder the successful execution of retrofit projects and impact their feasibility and effectiveness. Factors such as site-specific considerations, existing utilities, structures, wetlands, maintenance access, and land utilization play crucial roles in determining the feasibility and sustainability of a stormwater retrofit project. By carefully evaluating these considerations, municipalities can make informed decisions and ensure effective and responsible site utilization. Existing utilities, including water, electricity, and sewage systems, must be assessed to determine if they can adequately support the proposed development.

Existing drainage infrastructures warrant evaluation for their condition and potential for integration into retrofitting. Wetlands, as valuable ecosystems, require careful consideration to minimize any adverse impact on their ecological balance. Planning for maintenance access, such as roads and pathways, is essential for efficient site upkeep. Land utilization should be optimized to maximize the use of available space, taking into account factors such as zoning regulations and environmental impact. A careful analysis of these factors ensures that the project design and implementation align with the surrounding environment and infrastructure, a crucial aspect of the project's long-term success and sustainability. Table 4.2 presents a comprehensive list of key factors influencing the suitability of stormwater retrofits for a specific location.

Table 4.2 Stormwater retrofit site assessment

Factors	Considerations
Retrofit objectives	What is the primary goal driving the effort? Is it to manage stormwater effectively, focusing on controlling its quantity, improving its quality, or both?
Cost	Do stormwater retrofits show economic viability when evaluating the expected benefits against the initial investment and ongoing maintenance expenses?
Construction/ maintenance access	It is important to clearly define maintenance responsibilities for retrofits to avoid confusion or ambiguity regarding who is responsible for the upkeep and maintenance of the retrofitted elements.
Geological conditions	An analysis of the subsurface conditions at the site and their correlation with the specifications of the proposed retrofit is necessary. This includes factors such as soil permeability and groundwater depth. It is crucial to determine whether these conditions align with the proposed project objective, such as infiltration or noninfiltration-based best management practices.
Utilities	When considering the proposed retrofits, it is important to assess whether any conflicts exist between the current utilities and the intended locations. Such conflicts may necessitate either utility relocation or retrofit design modifications.
Contradictory land uses	It is important to assess whether retrofits align with the land uses of adjacent properties. This ensures that any property modifications or improvements harmonize with the surrounding area.
Ecological features, environmental aspects, and public safety	A thorough investigation of the potential effects of retrofits on nearby ecological features, such as environmentally sensitive bodies of water and vegetation, is imperative. Whenever technically and practically possible, measures must be identified and implemented to minimize or mitigate any adverse impacts. Is there a potential risk to public health and safety as a result of the retrofit?
Regulatory authorizations and certifications	Which regulatory bodies, at the local, state, or federal level, are responsible for overseeing the retrofit project, and can the retrofits receive approval from these entities?

When evaluating the effectiveness of retrofitted facilities in reducing pollutant loads, it is crucial to understand that their efficacy may not match that of newly designed and constructed facilities. However, a significant number of cases show potential for improvements in stormwater quantity and quality control. This observation holds true when considering the addition of a new goal for a preexisting project or the enhancement and expansion of an ongoing storm drainage system.

Achieving the desired effectiveness level might necessitate the incorporation of multiple small-scale practices, such as the implementation of green infrastructure practices. The issue of stormwater quantity often results in significant impacts on receiving waters and wetlands, primarily due to channel erosion. Therefore, it is essential for stormwater management facilities to maintain their ability to control stormwater quantity while also enhancing their pollutant removal effectiveness. Common stormwater retrofit applications for existing development and redevelopment projects involve implementing various measures. These measures include modifying drainage systems, installing stormwater management facilities, implementing stormwater controls at storm drain outfalls, and upgrading road culverts, rights-of-way, and parking lots to improve stormwater management. Additionally, they involve adopting in-stream practices within current drainage channels and establishing and rehabilitating wetlands.

4.5 CHAPTER SUMMARY

The complex interplay of inherent unpredictability and the looming threat of climate change may render current design standards for urban drainage systems obsolete. Over recent decades, the methodologies and assumptions of these processes have been rigorously examined due to the use of historical data that may be inaccurate or inconsistent. Initial urban drainage systems operated under the assumption of a relatively stable environment, with extrapolation of historical patterns and statistical analyses providing insights into future conditions. Climate change is expected to increase the unpredictability and intensity of extreme events, thus challenging established norms and standards for storm design.

Given the growing concern surrounding the requirements and benchmarks for urban drainage system design, there is a noticeable lack of a definitive path to address the typical design challenges presented by extraordinary events in an era characterized by dynamic and complex conditions. To further advance the conceptualization and establishment of design principles and benchmarks for urban drainage systems in the context of the "New Normal," this chapter provided an in-depth analysis of the methodology used in contemporary storm drainage infrastructure design. It explored various obstacles encountered in this process and potential additional measures and modifications, such as the impact of climate change, increased complexity, and the interaction between infrastructure and social systems.

Urban drainage infrastructure is significantly influenced by various physical factors. The sizing of components within the stormwater drainage system is crucial for facilitating the conveyance of the designated flow within a critical urban infrastructure network. Determining the appropriate level of service for a drainage infrastructure system, specifically in terms of the consequences of flooding and the acceptable frequency of such events, plays a vital role in establishing the magnitude of the design flow. This design flow is typically characterized as either a probability of flow exceedance or an interval between events of a similar nature. Upon determining the level of service and the frequency of design, the technical design of urban drainage systems employs approved methodologies, hydrologic input, and design parameters. The strategic allocation of resources to urban drainage infrastructure planning primarily focuses on the judicious selection of land and easements, as well as the effective management of their interaction with adjacent development.

Design flow magnitudes can be generated through the use of empirical peak discharge, hydrologic modeling, and statistical approaches based on hydrometric records. The prevalence of empirical peak runoff methods and hydrologic simulation models in urban drainage design can be attributed to their inherent advantage in determining design flow magnitudes within natural watercourses through the use of statistical methodologies. These two approaches facilitate the incorporation of spatial and temporal variability of site conditions by considering the land-use characteristics upstream of the infrastructure under construction, as well as the localized values of rainfall and snowmelt. Historically significant design storms represent the highest intensity of storm events that have occurred in a given area, while synthetic design storms refer to the combination of precipitation data averaged over time for a specific location.

The aforementioned methodologies for selecting rainfall and snowmelt design data depend on historical climatic data, assuming a static climate throughout the project's duration. This assumption, suggesting that past conditions can predict future ones, is invalidated by climate change. Specifically, the previously assumed constancy of design rainfall and snowmelt can no longer be maintained. Given that storm drainage infrastructure planning depends on factors like size and location, overlooking this fundamental concept could significantly impact design and strategic organization.

The traditional approach to designing and planning urban drainage systems has not accounted for the dynamic nature of precipitation patterns and hydro-climatic shifts. Consequently, historical IDF curves are not adjusted adequately to protect the public's immediate and long-term financial, environmental, and societal interests. Current storm design standards may be insufficient in the face of increasing climatic threats, such as unpredictable episodes of intense precipitation.

Ignoring the principle of nonstationarity can lead to concerns about climate variability, storm standards, and infrastructure design. This principle suggests that systems fluctuate across a continually shifting spectrum of variability. Previous procedures for estimating design rainfall intensities have shown that models used to build urban drainage infrastructure systems based on stationary rainfall data may not always be accurate. This is because they fail to take into account the non-stationarity nature of rainfall data, which is critical for accurately forecasting extreme weather rainfall events.

The complex nature of urban and infrastructural systems, combined with unpredictable climate variations, presents significant challenges to existing design and implementation paradigms. Urban and infrastructural systems demonstrate deep integration with social, ecological, and technical systems, accommodating numerous users and administrators. The integration and interdependence within and among systems often leads to increased complexity, making it difficult to identify or manipulate causal relations between elements, and the system's behavior and performance appear as emergent properties. The growing complexity and nonstationarity of urban and infrastructural systems pose significant challenges to risk analysis and management. The likelihood and unpredictability of disruptions increase when the causal linkages between events become unclear.

Implementing stringent design standards based on extreme weather conditions requires collaboration among planners, designers, and stakeholders, compelling them to engage in rigorous discussions and make difficult decisions. Using interdisciplinary, multi-hazard, and stakeholder approaches is considered the most effective strategy for adapting stormwater infrastructure system design criteria to the challenges posed by climate change and the issue's inherent complexity. Stakeholder engagement includes the fundamental principles of organizational coherence and transparency. Laws, formal agreements, and informal agreements serve as the regulatory frameworks governing storm drainage infrastructure standards across various organizations. The presence of above-ground flows adds a new level of complexity. These fundamental principles can provide a framework for policy formulation and implementation.

Climate change is expected to increase the frequency of extreme weather events significantly worldwide. To protect the well-being and security of the public, communities must proactively prepare for severe climatic conditions. While instrumental adaptation measures are primarily implemented at the local level, they must be integrated into a comprehensive public policy framework aimed at enhancing overall community resilience. This integration requires coordinated efforts across all government levels. These objectives, principles, and mechanisms act as catalysts for local climate adaptation policies while also providing an intergovernmental framework. The shift toward the "New Normal" of increased meteorological phenomena due to climate change necessitates a reevaluation of the criteria governing stormwater infrastructure design and drainage systems. The following guidelines are suggested for optimal performance:

1. Incorporate elements of uncertainty, complexity, and adaptation into the storm criteria and risk assessment framework for urban infrastructure. Return-period and threshold-based methodologies can improve certain components and subsystems, although they may require more frequent updates. Safety considerations should be incorporated, and the standards for infrastructure design and construction thresholds should be elevated, with a focus on accounting for climate change and its associated variability.

2. Define and adapt procedures at the system level. The integration of nature-based solutions and green infrastructure can enhance urban planning that primarily focuses on infrastructure development. Risk analysis and reasoning can help balance the concepts of *safe-to-fail* and *fail-safe*, acknowledging the inevitability of disruptions. Prioritizing individuals should take precedence over considerations related to infrastructure, hazards, provisions, and consequences. The use of experimental capacity, continuous performance evaluation, and ongoing development of standards and procedures should all exceed the limitations imposed by static design requirements. Understanding urban and design concepts should be expanded to include the complex interplay among environmental, social, and technical complexities, thereby promoting the development of multifaceted solutions rather than purely technical ones.

4.6 CHAPTER PROBLEMS

1. The nonstationary probability distribution information for GEV model parameters is as follows: $uo = 250.65$, $u1 = 0.1574$, $\sigma = 95.7$, and $k = 0.04565$. Determine the location parameter

for the specified time. Utilize GEV parameters to ascertain the discharge corresponding to a 20-year return period at $t = 0$ and $t = 10$ for return periods of 5, 10, and 50 years.

2. Download 5-year Daily Precipitation Data from the National Oceanic and Atmospheric Administration (NOAA) Precipitation Frequency Data Server (PFDS) (start at https://www.ncei.noaa.gov/maps/daily-summaries/) and perform a space-time nonstationary extreme value analysis on the winter 3-day maximum precipitation.

3. Fit a spatial model to each of the coefficients and to the following scale and shape parameters. For a few wet and dry years (which you can select based on the average spatial three-day precipitation), obtain the 2-year, 50-year, and 100-year return levels on the spatial grid. Using spatial models, obtain the GEV parameters at each grid point. Estimate the 2-year, 50-year, and 100-year return levels at each grid point and map them. Compare the 2-year return levels with the observed values in the selected years. For a couple of representative locations, plot the time series of the three-day precipitation maximum along with the time-varying return levels. Compare them with the stationary return levels.

4. The frequency and severity of extreme climatic events are on the rise, which raises concerns about the preparedness of our infrastructure to handle these shifts. Current infrastructure design heavily relies on IDF curves, which assume that extreme weather events will remain relatively constant over time. However, climate change is expected to alter climatic extremes—a phenomenon referred to as nonstationarity. Discuss and demonstrate how the current IDF curves can significantly underestimate the magnitude of extreme precipitation events when nonstationarity is present. Thus, relying on these curves for infrastructure design in a changing climate may not be suitable.

5. Use the ground-based observations of annual maximum extreme precipitation data from the U. S. National Oceanic

Atmospheric Administration (NOAA) Atlas for JFK International Airport, spanning from January 1980 to January 2022, to answer the following questions:

a. Construct stationary IDF curves for each station using GEV distribution. Disaggregate annual maximum daily rainfall data and calculate stationary rainfall intensities for each station. Determine the maximum intensities for durations ranging from 1 to 24 hours and return periods from 2 to 100 years. This analysis will address the following questions using the NOAA ground-based observations of annual maximum extreme precipitation data for JFK International Airport.

b. Generate IDF curves for different return periods (2, 10, and 100 years) under both stationary and nonstationary conditions.

c. Illustrate the potential risks associated with floods and infrastructure failures that can result from the underestimation of extreme precipitation events due to the adoption of a stationary climate assumption.

d. Is it possible to transform the annual maxima rainfall data, derived from daily time series, into nonstationary GEV distributions?

e. Briefly explain how the integration of nonstationary IDF curves in future urban drainage design concepts addresses the need to adapt urban drainage systems to changing climatic conditions.

f. Demonstrate that a higher degree of nonstationary behavior is observed relative to long durations with short return periods.

6. Consider retrofitting the current drainage systems for a 5-acre plot of land with a corresponding drainage area and a proposed impervious cover of 80% for a commercial development project. The proposed approach involves implementing an infiltration-based retention pond to address water quality

problems, given a filter bed depth of 5 feet and a maximum ponding depth of 6 inches:

    a. What is the definition of water quality volume?

    b. How does one calculate the required water quality volume and surface area?

    c. Given the Environmental Protection Agency cost estimate formula (Cost Estimate = 7.30 * V^0.99), what is the cost of implementing an infiltration-based retention pond?

7. Define and explain the three major components of low-impact development design.

8. What are the objectives and benefits of stormwater retrofits?

9. Discuss common permeable pavement applications, types, and suitable applications for the retrofit of storm drainage systems.

10. Mini-group project opportunities: Conduct a comprehensive analysis of specific regions with documented concerns related to water quality or flooding. This analysis will enable the identification of retrofit options that offer the highest level of cost-effectiveness in addressing these issues. Provide a proposal that demonstrates the selected project domain, potentially involving a cooperative endeavor with various stakeholders:

    a. Define the retrofit project goal.

    b. Determine the stormwater control measures (SCMs) best suited for the site.

    c. Evaluate different retrofit scenarios for the stormwater drainage system if the goal is to mitigate the potential risk of flooding. Calculate the resilience and reduction of an SCM's failure to extreme storm events.

11. The "New Normal" municipality has proposed a comprehensive watershed management plan with land conservation as a key component. Land conservation is crucial for improving

the efficacy of stormwater management methods by encouraging sustainable land use and setting long-term restrictions on land use. Consequently, the municipality employs consultants to devise strategies to achieve the intended goal. As a consulting engineer, provide a detailed description and explanation of the watershed management plan that the city can implement. This comprehensive plan will include a variety of regulatory measures aimed at managing the city's watershed effectively.

SELECTED SOURCES AND REFERENCES

Achleitner, S., M. Möderl, and W. Rauch. 2007. "CITY DRAIN©— An Open Source Approach for Simulation of Integrated Urban Drainage Systems." *Environmental Modelling & Software* 22, no. 8: 1184–1195.

Ahmed, A. U. 2005. "Adaptation Options for Managing Water Related Extreme Events Under Climate Change Regime: Bangladesh Perspectives." In *Climate Change and Water Resources in South Asia*. Leiden: Balkema Press, pp. 255–278.

Araya, F. and S. Vasquez. 2022. "Challenges, Drivers, and Benefits to Integrated Infrastructure Management of Water, Wastewater, Stormwater and Transportation Systems." *Sustainable Cities and Society* 82: 103913.

Arnbjerg-Nielsen, K. and H. S. Fleischer. 2009. "Feasible Adaptation Strategies for Increased Risk of Flooding in Cities Due to Climate Change." *Water Science and Technology* 60, no. 2: 273–281.

Asquith, W. H. and J. S. Famiglietti. 2000. "Precipitation Areal-Reduction Factor Estimation Using an Annual-Maxima Centered Approach." *Journal of Hydrology*, 230: 55–69. doi:10.1016/S0022-1694(00)00170-0.

Beven, K. J. 2011. "I Believe in Climate Change but How Precautionary Do We Need to Be in Planning for the Future?" *Hydrological Processes* 25: 1517–1520. doi:10.1002/hyp.7939.

Bras, R. L. and I. Rodriguez-Iturbe. 1993. *Random Functions and Hydrology*. Massachusetts: Addison-Wesley.

Burian, S. J. and F. G. Edwards. 2002. "Historical Perspectives of Urban Drainage." In *Global Solutions for Urban Drainage*, pp. 1–16.

Butler, D., C. Digman, C. Makropoulos, and J. W. Davies. 2018. *Urban Drainage*. CRC Press.

Chow, W. T. 2018. "The Impact of Weather Extremes on Urban Resilience to Hydro-Climate Hazards: A Singapore Case Study." *International Journal of Water Resources Development* 34, no. 4: 510–524.

De Michele, C., N. T. Kottegoda, and R. Rosso. 2001. The Derivation of Areal Reduction Factor of Storm Rainfall from Its Scaling Properties. *Water Resources Research*, 37 (12): 3247–3252. doi:10.1029/2001WR000346.

Djordjević, S., D. Butler, P. Gourbesville, O. Mark, and E. Pasche. 2011. "New Policies to Deal with Climate Change and Other Drivers Impacting on Resilience to Flooding in Urban Areas: The CORFU Approach." *Environmental Science & Policy* 14, no. 7: 864–873.

Dong, X., H. Guo, and S. Zeng. 2017. "Enhancing Future Resilience in Urban Drainage System: Green Versus Grey Infrastructure." *Water Research* 124: 280–289.

Elliott, A. H. and S. A. Trowsdale. 2007. "A Review of Models for Low Impact Urban Stormwater Drainage." *Environmental Modelling & Software* 22, no. 3: 394–405.

Ellis, B., C. Viavattene, M. Revitt, C. Peters, and H. Seiker. 2009. "A Modelling Approach to Support the Management of Flood and Pollution Risks for Extreme Events in Urban Stormwater Drainage Systems." In *4th Switch Scientific Meeting*, pp. 4–7.

Haghighatafshar, S., P. Becker, S. Moddemeyer, A. Persson, J. Sörensen, H. Aspegren, and K. Jönsson. 2020. "Paradigm Shift in Engineering of Pluvial Floods: From Historical Recurrence Intervals to Risk-Based Design for an Uncertain Future." *Sustainable Cities and Society* 61: 102317.

Hein, D. K. and P. Eng. 2015. "Maintenance Guidelines for Permeable Interlocking Concrete Pavement Systems." In *International Conference on Concrete Block Pavements*.

Jenkins, K., S. Surminski, J. Hall, and F. Crick. 2017. "Assessing Surface Water Flood Risk and Management Strategies Under Future Climate Change: Insights from an Agent-Based Model." *Science of the Total Environment* 595: 159–168.

Kang, N., S. Kim, Y. Kim, H. Noh, S. J. Hong, and H. S. Kim. 2016. "Urban Drainage System Improvement for Climate Change Adaptation." *Water* 8, no. 7: 268.

Keifer, C. J. and H. H. Chu. 1957. Synthetic Storm Pattern for Drainage Design. Proceedings ASCE, *Journal of the Hydraulics Division*, 83 (4): 1–25.

Kim, Y., D. A. Eisenberg, E. N. Bondank, M. V. Chester, G. Mascaro, and B. S. Underwood. 2017. "Fail-Safe and Safe-to-Fail Adaptation: Decision-Making for Urban Flooding Under Climate Change." *Climatic Change* 145, no. 3: 397–412.

Kirshen, P., L. Caputo, R. M. Vogel, P. Mathisen, A. Rosner, and T. Renaud. 2015. "Adapting Urban Infrastructure to Climate Change: A Drainage Case Study." *Journal of Water Resources Planning and Management* 141, no. 4: 04014064.

Koutsoyiannis, D., A. Montanari, H. F. Lins, and T. A. Cohn. 2009. "Climate, Hydrology and Freshwater: Towards an Interactive Incorporation of Hydrological Experience into Climate Research." Discussion of "The Implications of Projected Climate Change for Freshwater Resources and Their Management."

Kron, W., J. Eichner, and Z. W. Kundzewicz. 2019. "Reduction of Flood Risk in Europe—Reflections from a Reinsurance Perspective." *Journal of Hydrology* 576: 197–209.

Krysanova, V., Ch. Donnelly, A. Gelfan, D. Gerten, B. Arheimer, F. Hattermann, and Z. W. Kundzewicz. 2018. "How the Performance of Hydrological Models Relates to Credibility of Projections Under Climate Change." *Hydrological Sciences Journal* 63, no. 5: 696–720. doi:10.1080/02626667.2018.1446214.

Kundzewicz, et al. 2008. *Hydrological Sciences Journal* 54, no. 2: 394–405. doi:10.1623/hysj.54.2.394.

Kundzewicz, Z. W., J. Huang, I. Pinskwar, B. Su, M. Szwed, and T. Jiang. 2020. "Climate Variability and Floods in China—A Review." *Earth-Science Reviews*, 103434. https://doi.org/10.1016/j.earscirev.2020.103434.

Madsen, H., K. Arnbjerg-Nielsen, and P. S. Mikkelsen. 2009. "Update of Regional Intensity–Duration–Frequency Curves in Denmark: Tendency Towards Increased Storm Intensities." *Atmospheric Research* 92, no. 3: 343–349.

Madsen, H., D. Lawrence, M. Lang, M. Martinkova, and T. R. Kjeldsen. 2014. "Review of Trend Analysis and Climate Change Projections of Extreme Precipitation and Floods in Europe." *Journal of Hydrology* 519: 3634–3650. doi:10.1016/j.jhydrol.2014.11.003.

Mailhot, A. and S. Duchesne. 2010. "Design Criteria of Urban Drainage Infrastructures Under Climate Change." *Journal of Water Resources Planning and Management* 136, no. 2: 201–208.

Meyer, M. D. and B. Weigel. 2011. "Climate Change and Transportation Engineering: Preparing for a Sustainable Future." *Development* 1943, no. 5436: 0000108.

Miller, J. D. and M. Hutchins. 2017. "The Impacts of Urbanisation and Climate Change on Urban Flooding and Urban Water Quality: A Review of the Evidence Concerning the United Kingdom." *Journal of Hydrology: Regional Studies* 12: 345–362.

Mills, F., J. Kohlitz, N. Carrard, and J. Willetts. 2019. "Considering Climate Change in Urban Sanitation: Conceptual Approaches and Practical Implications." USHHD Learning Brief. The Hague: SNV.

Mugume, S. N. and D. Butler. 2017. "Evaluation of Functional Resilience in Urban Drainage and Flood Management Systems Using a Global Analysis Approach." *Urban Water Journal* 14, no. 7: 727–736.

NERC. Natural Environmental Research Council. 1975. Flood studies report, vol. II, Meteorological Studies, Swindon, England.

NERC. 1977. Flood studies supplementary report No 1: The areal reduction factor in rainfall frequency estimation. Natural Environment Research Council, UK.

Neumann, J. E., J. Price, P. Chinowsky, L. Wright, L. Ludwig, R. Streeter, and J. Martinich. 2015. "Climate Change Risks to U.S. Infrastructure: Impacts on Roads, Bridges, Coastal Development, and Urban Drainage." *Climatic Change* 131, no. 1: 97–109.

Rodrigues-Iturbe, I. and J. M. Mejia. 1974. "On the Transformation of Point Rainfall to Areal Rainfall." *Water Resources Research*, 10(4): 729–735. doi:10.1029/WR010i004p00729.

Salimi, M. and S. G. Al-Ghamdi. 2020. "Climate Change Impacts on Critical Urban Infrastructure and Urban Resiliency Strategies for the Middle East." *Sustainable Cities and Society* 54: 101948.

Thakali, R., A. Kalra, and S. Ahmad. 2016. "Understanding the Effects of Climate Change on Urban Stormwater Infrastructures in the Las Vegas Valley." *Hydrology* 3, no. 4: 34.

Thakali, R., A. Kalra, S. Ahmad, and K. Qaiser. 2018. "Management of an Urban Stormwater System Using Projected Future Scenarios of Climate Models: A Watershed-Based Modeling Approach." *Open Water* 5, no. 2: 1.

Uyumaz, A. 1994. "Highway Storm Drainage with Kerb-Opening Inlets." *Science of the Total Environment* 146: 471–478.

van der Pol, T. D., E. C. van Ierland, and S. Gabbert. 2017. "Economic Analysis of Adaptive Strategies for Flood Risk Management Under Climate Change." *Mitigation and Adaptation Strategies for Global Change* 22, no. 2: 267–285.

Walesh, S. G. 1991. *Urban Surface Water Management*. John Wiley & Sons.

Wang, M., Y. Zhang, A. E. Bakhshipour, M. Liu, Q. Rao, and Z. Lu. 2022. "Designing Coupled LID–GREI Urban Drainage Systems: Resilience Assessment and Decision-Making Framework." *Science of The Total Environment* 834: 155267.

Willems, P. 2013. "Revision of Urban Drainage Design Rules After Assessment of Climate Change Impacts on Precipitation Extremes at Uccle, Belgium." *Journal of Hydrology* 496: 166–177.

Willems, P. and J. Olsson. 2012. "Impacts of Climate Change on Rainfall Extremes and Urban Drainage Systems." IWA Publishing.

Willems, P., K. Arnbjerg-Nielsen, J. Olsson, and V. T. V. Nguyen. 2012. "Climate Change Impact Assessment on Urban Rainfall Extremes and Urban Drainage: Methods and Shortcomings." *Atmospheric Research* 103: 106–118.

Winter, K. and S. Mgese. 2011. "Stormwater Drainage: A Convenient Conduit for the Discharge of Urban Effluent into the Berg River, South Africa." In *12th International Conference on Urban Drainage*, Porto Alegre, Brazil.

Wright, G. B., L. B. Jack, and J. A. Swaffield. 2006. "Investigation and Numerical Modelling of Roof Drainage Systems Under Extreme Events." *Building and Environment* 41, no. 2: 126–135.

Wytyczne obliczania ilosci wód opadowych i roztopowych na obszarze kolejowym Is-2 [Guidelines for calculating the amount of rainwater and meltwater in the Is-2 railway area]. 2019. PKP Polskie Linie Kolejowe S.A., Warsaw.

Wyzga, B., Z. W. Kundzewicz, R. Konieczny, M. Piniewski, J. Zawiejska, and A. Radecki-Pawlik. 2018. "Comprehensive Approach to the Reduction of River Flood Risk: Case Study of the Upper Vistula Basin." *Science of the Total Environment* 631–632: 1251–1267. https://doi.org/10.1016/j.scitotenv.2018.03.015.

Yazdanfar, Z. and A. Sharma. 2015. "Urban Drainage System Planning and Design—Challenges with Climate Change and Urbanization: A Review." *Water Science and Technology* 72, no. 2: 165–179.

Zhou, Q. 2014. "A Review of Sustainable Urban Drainage Systems Considering the Climate Change and Urbanization Impacts." *Water* 6, no. 4: 976–992.

Zhou, Q., G. Leng, J. Su, and Y. Ren. 2019. "Comparison of Urbanization and Climate Change Impacts on Urban Flood Volumes: Importance of Urban Planning and Drainage Adaptation." *Science of the Total Environment* 658: 24–33.

Zhou, Q., T. E. Panduro, B. J. Thorsen, and K. Arnbjerg-Nielsen. 2013. "Adaption to Extreme Rainfall with Open Urban Drainage System: An Integrated Hydrological Cost-Benefit Analysis." *Environmental Management* 51, no. 3: 586–601.

5

SUPPLEMENTING URBAN DRAINAGE SYSTEM DESIGN CRITERIA TO ACCOMMODATE THE "NEW NORMAL"

5.1 INTRODUCTION

Climate-related issues such as sea-level rise, extreme weather, and eco-system disruption have become increasingly prevalent worldwide over recent decades. Cities with extensive infrastructure face new threats due to weather nonstationarity. Urban infrastructure, built in recent decades, is designed to provide enduring service and withstand antici-pated disruptions based on historical data. Despite the meticulous risk estimation required by engineering standards, the unpredictable na-ture of weather events, exacerbated by climate change, poses increas-ing challenges.

Extreme weather events have led to numerous catastrophic infra-structure failures. Table 5.1 lists extreme weather events worldwide from 1980 to 2021. The inability to anticipate the intensity and impact of hurricanes can be attributed to the inadequate resilience of infra-structure systems that were not designed to withstand such events. However, these weather phenomena were somewhat predictable based on the geographic features and historical meteorological data of the affected regions. Infrastructure failures have had unanticipated con-sequences, including unexpected human casualties, property damage,

Table 5.1 Extreme weather events worldwide, 1980–2021
Source: The National Oceanic and Atmospheric Administration and
Emergency Events Database

Event Type	Year	Country	Damage (USD)	Deaths
Hurricane Katrina	2005	USA (LA/MS/AL/FL)	180 billion	>1,085
Hurricane Harvey	2017	USA (LA/TX)	141 billion	89
Hurricane Maria	2017	USA (PR/VI)	102 billion	2,981
Hurricane Sandy	2012	USA (NY/NJ/CT)	80 billion	159
Hurricane Ida	2021	USA (LA/MS/NJ/NY/CT)	75 billion	96
Hurricane Irma	2017	USA (FL/GA/SC/PR)	56 billion	97
Hurricane Andrew	1992	USA (FL/LA)	54 billion	62
Flooding	1988	China	51 billion	3,656
Flooding	2011	Thailand	49 billion	813
Hurricane Ike	2008	USA (TX/LA/MS)	40 billion	112
Hurricane Wilma	2005	Cuba, Mexico, USA (FL)	36 billion	35
Hurricane Ivan	2004	USA (FL/LA)	31 billion	57
Flooding	2021	China	30 billion	347
Hurricane Michael	2018	USA (FL/GA)	28 billion	49
Flooding	1995	North Korea	27 billion	68
Hurricane Rita	2005	USA (LA/TX)	27 billion	119
Hurricane Florence	2018	USA (NC/SC)	26 billion	53
Hurricane Ivan	2004	USA	26 billion	123
Flooding	2016	China	26 billion	475
Hurricane Charley	2004	USA (FL)	24 billion	35
Flooding	2010	China	23 billion	1,691
Flooding	2021	Germany and Belgium	22 billion	240
Flooding	1996	China	22 billion	2,775
Hurricane Laura	2020	USA (TX/LA/MS/AR)	20 billion	42

loss of public services, disruption of vital infrastructure, interruption of businesses and livelihoods, health risks, environmental losses, and negative impacts on regional economies.

The current approach to infrastructure development involves identifying potential risks to ensure that urban infrastructure systems are reliable and maintain their functionality up to a designed system capacity, commonly referred to as *fail-safe* (see Figure 5.1). Recent advancements in climate prediction models have facilitated a more comprehensive understanding and forecasting of the potential implications of extreme weather events on infrastructure. By projecting future climate scenarios,

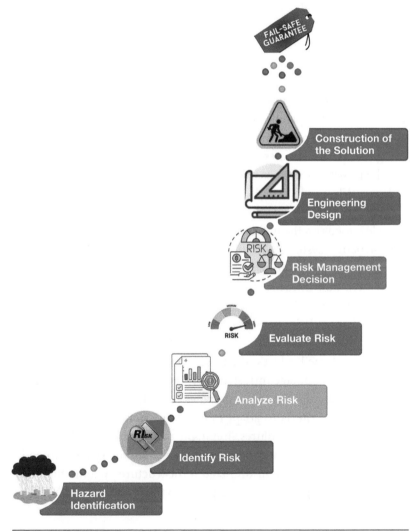

Figure 5.1   Current fail-safe infrastructure development process that incorporates probabilistic risk calculation

professionals can evaluate the range of climatic risks that may be encountered. Several factors contribute significantly to the uncertainties in weather predictions, including the complex nature of weather phenomena, greenhouse gas (GHG) emissions, population dynamics, land-use patterns, technological advancements, and natural resource utilization.

The unpredictable nature of weather phenomena, influenced by numerous variables and the dynamic nature of future climate change, can pose challenges for engineered infrastructure in managing associated risks effectively. A fresh perspective on infrastructure development is needed, one that transcends the traditional risk assessment approach and focuses on managing the consequences of unforeseen risks that may exceed the capabilities of engineered systems.

Lessons from recent hurricanes suggest the advisability of considering potential outcomes in unpredictable climate conditions, including the possibility of risks overwhelming infrastructure and leading to failures. While substantial effort has been dedicated to determining the ideal risk threshold for creating resilient infrastructure, there has been less focus on the potential negative repercussions on both critical infrastructure systems and the surrounding area should risks surpass the engineered threshold. In an ever-changing future, it is crucial to implement new infrastructure development practices to protect both critical infrastructure systems and their surrounding area, as well as vulnerable populations, from the potential dangers posed by extreme weather conditions and infrastructure failure.

This chapter will address the challenges that climate change poses to current urban drainage infrastructure systems, their design standards and guidelines, and the methodologies for enhancing urban drainage system design criteria to adapt to evolving circumstances. It aims to define infrastructure failures, highlight the complexities inherent in the planning and design process, and offer preliminary recommendations for designing urban drainage infrastructure systems that align with current design criteria in response to the "New Normal."

5.1.1 Infrastructure Within a Nonstationary Climate

Past infrastructure development practices face challenges in accommodating recent climate fluctuations. In the development process, engineers commonly use prepackaged datasets and charts to express weather-related hazards. These tools provide statistics on temperature, precipitation, wind speed, and other factors to help assess the severity of weather events, such as a 100-year event, and determine operational thresholds. However, given the current fluctuations in weather patterns, these models may not effectively aid infrastructure planning for potential risks in a constantly evolving and unpredictable environment. Effective planning and strategic decision making require consideration of a wide range of factors, including data and models developed by the scientific community regarding future climate projections. Modifying drainage design standards and the drainage infrastructures themselves can be challenging, even in the face of changes in governance, society, the economy, and ecological systems. Some infrastructure systems may be outdated in terms of risk management, thereby offering room for improvement.

Addressing the impact of unusual weather patterns necessitates the implementation of innovative infrastructure strategies for development, operation, and management. A fundamental element of engineering development is the ability to effectively predict or describe future circumstances to evaluate the consequences of design choices comprehensively. For instance, levees are typically designed to manage and provide protection against floods with a frequency of 100 years or more, aiming to ensure safety.

The conventional method for engineering design assumes that future events will resemble past occurrences and that a sample of observations from identical events can represent the past and predict the future, a concept known as *stationarity*. However, predicting the impact of changing climates on cities is becoming increasingly complex due to the various physical and natural processes involved. While cities have a proven track record of adapting to such changes through infrastructure development, relying solely on current models and data for conventional adaptation efforts may not provide adequate risk management solutions for the future. The expansion of urban areas has the potential to impact the quality and quantity of stormwater run-off discharge or volume. Incorporating *nonstationarity*, a concept that may have been overlooked in past development methodologies, is now essential for optimizing future infrastructure strategies.

The current scientific discourse emphasizes the use of both stationary and nonstationary methodologies in predicting the frequency and intensity of future climate extremes. This underscores the importance of implementing infrastructure practices that can adapt effectively to future weather extremes. The integration of high-resolution stationary models could enhance the depiction of the "New Normal" or extreme weather phenomena in specific regions. Indeed, stationary models may offer advantages over nonstationary models, potentially improving the accuracy of predicted extreme frequencies and providing a means of assessing uncertainty. Concurrently, practitioners are showing increasing interest in models that consider nonstationary trends in extreme frequency analysis, achieved by including climatic covariates such as time and temperature. Nonstationary models, such as generalized extreme value models, may be more suitable for representing extremes than stationary models, as various studies have shown.

Predicting extreme weather events has become increasingly challenging because of urbanization and human-induced environmental changes, regardless of the specific model used for analysis. While it has been common to rely on statistical and frequency-based data to establish historical practices, these methods may not always provide accurate predictions for inherently unpredictable future events. It is crucial to acknowledge the inherent uncertainty associated with future potential risks due to climate change. Therefore, it is imperative to adopt strategic decision-making processes that incorporate insights from climate experts, legislators, and professionals. This ensures that infrastructure development practices are well-informed and capable of effectively addressing the challenges posed by unpredictable potential risks. City municipalities must implement a robust infrastructure risk management approach to safeguard urban drainage infrastructure and ensure its smooth operation.

5.1.2 Understanding Urban Drainage Infrastructure Failures

In urban drainage infrastructure planning, *failure* is primarily considered in the context of preventive measures. Current infrastructure development practices, often referred to as *fail-safe*, emphasize minimizing the occurrence of failures by adhering to and upholding plans and designs. Urban drainage infrastructure failure can be categorized into two distinct cases. Type-1 failures occur when a system or process fails to execute its designated function, leading to adverse outcomes. Type-2 failures, on the other hand, occur when potential risks emerge and disrupt preexisting infrastructure. It is important to acknowledge that these disruptions can have substantial ramifications on multiple levels, including on the economy, environment, and surrounding communities.

The potential consequences of these failures can be significantly detrimental, affecting the physical and mental health of individuals, the operational efficiency of businesses, and the overall equilibrium of ecological systems. Type-1 failures often occur when the infrastructure lacks the capacity to effectively handle or reduce anticipated risks or when it fails to fulfill its intended purpose. A notable example of this type of failure involves the transportation of an excessive amount of stormwater runoff through drainage structures that exceeds their design capacity. This results in the discharge of untreated or partially treated sewage, or it may lead to water quality issues if the drainage system is a combined sewer. Type-2 failures occur when the consequences of a Type-1 failure exceed the operational capabilities of the infrastructure, leading to critical damage on multiple levels, including the system's integrity, extensive water inundation, property destruction, ecological catastrophes, financial losses, and impacts on livelihoods. The fail-safe approach is a risk management strategy that prioritizes the reduction or prevention of Type-1 failures.

The *safe-to-fail* approach integrates Type-2 failures and assesses potential risks. It is important to highlight scenarios in which Type-2 failures occur without understanding the cause or prognosis of Type-1 failures. The occurrence of Type-1 failures can often be attributed to unanticipated risks, underscoring the importance of addressing this issue. It is crucial to acknowledge that the occurrence of fail-safe infrastructure and subsequent Type-2 failure often stems from the underlying causes of its catastrophic breakdowns. These failures are not a result of insufficient data on potential risks but rather a lack of consideration for the consequences that arise from infrastructure failure. The concept of infrastructure failures encompasses the mitigation of potential future climate risks by prioritizing an understanding of the consequences resulting from the loss of infrastructure services rather than attempting to predict the improbable factors leading to failure. Understanding the potential effects of infrastructure failure is important, emphasizing the connections between the loss of infrastructure services and their resulting consequences.

Table 5.2 provides a summary of infrastructure service instances and their corresponding Type-1 and Type-2 failures. The safe-to-fail infrastructure failure methodology is utilized through the analysis of historical data, such as occurrences of drainage infrastructure component failures (e.g., type and size of pipe, inlet catch basin), frequency of maintenance, and physical breakdowns. These data are often documented in the operations and maintenance logs of municipalities and serve as valuable resources for improving the planning and design practices of urban drainage infrastructure.

Failures in drainage infrastructure can be attributed to deficiencies in long-term strategic asset management development. These deficiencies arise from a perspective that views infrastructure as a collection of individual components with separate functions rather than recognizing them as interconnected systems operating within the larger watershed scale. This perspective overlooks the disruptions caused by the loss of these services and the potential consequences, both intended and unintended, that may result from infrastructure failure.

Table 5.2 Understanding drainage infrastructure failures

Type 1 Failure	Type 2 Failure
1. Overcapacity/overflow	1. Damage or loss incurred to physical assets and essential services
2. Levy failure	2. Water source contamination
3. Water quality and/or quantity control problems	3. Water entering and permeating through a substance or medium
4. Over-capacity stormwater, sewer pipe	4. Decrease in revenue or decline in commercial activity
5. Sump pump or pump failure or collapse	5. Waterborne diseases via harmful microorganisms or pathogens
6. Surface inflow/overland flooding	6. Issues with coastal vegetation and ecosystems
7. Drainage foundation failure	7. Potential risks and vulnerabilities impacting trade, commerce, and tourism
8. Storm surge gates and flood barriers failure	8. Propagation of failures across multiple systems given the interdependencies between various utility and infrastructure systems
9. Destruction of streets, highways, and railroads	9. Impact of mobility regarding the effects and consequences of the ability to move or be moved freely and easily, encompassing various aspects of modes of transportation
	10. Impairment of evacuation routes and emergency services

Drainage infrastructure failures can occur when the systems experience a disruption or a complete cessation of operation, irrespective of the cause. This is distinct from a malfunction in a specific component of the system. Consequently, it is essential to view drainage infrastructure systems as interconnected networks rather than as isolated components. By definition, the precise assessment of severe climate-related risks or infrastructure component losses is insignificant in the design of drainage infrastructure when considering Type-1 and Type-2 failures (see Figure 5.2). These failures are not dependent on the occurrence of initiating events. This definition facilitates the application of

Figure 5.2   Example of impairment of evacuation routes and emergency services Type-2 failure

both static and dynamic models to predict climatic risk in the practices of drainage infrastructure development.

5.2 FAIL-SAFE AND SAFE-TO-FAIL DRAINAGE SYSTEM DESIGN CONCEPTS

The concept of fail-safe and safe-to-fail design in drainage infrastructure is a crucial aspect of urban planning and engineering. These approaches seek to minimize potential risks and consequences associated with system failures while promoting adaptability and resilience under changing conditions. Steiner introduced *safe-to-fail* design principles in 2006, focusing on predicting and managing infrastructure failures. The central premise is to design systems strategically to contain and minimize potential failures. Recent extreme weather events have demonstrated that the failure of robust, fail-safe infrastructure systems can have catastrophic consequences, exacerbating the vulnerability of already susceptible populations and intensifying the overall impact of such events. There is growing concern that existing infrastructures, designed for robustness and resistance to operational and structural failures, may not be sufficient to withstand the impacts of "New Normal" nonstationarity due to climate change (Steiner 2006; Kim et al. 2017).

Safe-to-fail urban drainage infrastructure is a system deliberately designed to lose its operational capability in a controlled, intentional, and predictable manner. This allows for different types of consequences based on prioritized decisions, even when unforeseen risks exceed safety thresholds. The safe-to-fail approach enables the drainage network to operate in three distinct states: operational, partially operational or encountering failure, and complete failure with predefined outcomes. The normal state of the drainage network is defined by its ability to perform all intended functions within its designated capacity while operating within a predetermined range of potential risks.

Limited functionality or system failure, referred to as a Type-1 failure, occurs when a system can no longer provide its intended services. However, it mitigates the consequences of a Type-2 failure within the system. In safe-to-fail designed systems, a total failure can occur, resulting in the loss of system functionality. However, the consequences of a Type-1 failure are effectively mitigated to minimize the overall impact of a Type-2 failure. This includes reducing casualties, mitigating ecological impact, minimizing adverse economic consequences, and preventing damage to critical assets. These decisions are based on empirical knowledge and considerations related to software engineering and system development. To implement safe-to-fail development practices effectively, the drainage infrastructure network must have the necessary capacity to handle the repercussions of a complete failure efficiently. This can be achieved by restoring the failed operational component or retrofitting and integrating the drainage network with large-scale green infrastructure.

The safe-to-fail principle is crucial when addressing unforeseen risks arising from a nonstationary climate. Municipalities can effectively implement long-term climate adaptation measures using the safe-to-fail framework. By implementing such systems, cities, in collaboration with multiple stakeholders, are encouraged to evaluate their drainage system's capacity to handle unpredictable risks and develop flexible coping strategies. In the context of drainage infrastructure management, it is important to address frequent controlled failures to mitigate potential risks associated with unpredictable weather extremes. Proactively addressing these failures can prevent the adoption of risky development practices that may lead to detrimental consequences. Additionally, this approach emphasizes the importance of reassessing calculated risks and actively experiencing anticipated outcomes. This evaluation is crucial for effectively managing the impact of evolving climatic conditions.

During the implementation of a safe-to-fail technique, municipalities must thoroughly evaluate and anticipate the potential outcomes of both Type-1 and Type-2 failure scenarios during the planning and design phases. Incorporating diverse perspectives and expertise from multiple stakeholders can contribute to a comprehensive understanding and assessment of large-scale urban watershed areas, as well as the complex interdependencies among their drainage infrastructure systems. The integration of fail-safe infrastructure development optimizes the management of consequences in Type-1 failure scenarios due to its ability to enable strategic allocation of system resources and implementation of contingency measures.

For instance, stormwater retrofitting using green infrastructure, including designated buffer zones, green streets, large street medians, bioretention basins, parks, vegetated flood buffers, and golf courses, can be strategically implemented to handle the negative impacts of excessive runoff efficiently. When implementing drainage system retrofitting, it is important to consider the recovery capacity of nearby areas in relation to flooding due to extreme weather events. By considering these factors, stormwater retrofitting using green infrastructure can efficiently mitigate the impacts of excessive runoff. These approaches are considered safe-to-fail for Type-2 failures due to their ability to operate similarly to traditional stormwater conveyance systems. These design approaches aim to enhance current infrastructure development practices and focus on resilience. The development of these approaches involves the active participation of policymakers, planners, engineers, landscape architects, and various other stakeholders.

The safe-to-fail design approach recognizes the potential for unexpected severe weather conditions and prioritizes design choices that enhance the adaptive capacity and resilience of urban systems. Adaptive capacity refers to the ability to respond to both anticipated and unanticipated hazards by ensuring the provision of essential services. The concept of resilience aligns with the desired characteristics of drainage infrastructure designed to be safe in the event of failure. Transitioning from a fail-safe design approach to a safe-to-fail design approach requires a shift from traditional design approaches that focus on preventing individual component failure to a more holistic approach that includes:

1. Prioritizing the proper operation of critical drainage infrastructure services throughout the urban watershed over the prevention of minor drainage component failures
2. Concentrating on minimizing the effects of extreme weather events, such as peak stormwater runoff discharge or volume, rather than reducing the probability of damage
3. Implementing strategies that maintain and enhance the provision of social and ecosystem services
4. Creating decentralized and autonomous drainage infrastructure systems instead of centralized and hierarchical systems
5. Encouraging interdisciplinary communication and collaboration rather than merely integrating disparate disciplinary perspectives

Adopting this perspective also necessitates a broader range of decision-making criteria rather than risk-based approaches, including the preservation of ecosystem services, provision of social equity, facilitation of innovation, and enhancement of disaster response procedures. The implementation of fail-safe design strategies aims to enhance the resilience of drainage infrastructure to withstand increasingly extreme weather events. The concept of climate stationarity is a fundamental assumption used in the formulation of fail-safe strategies. However, safe-to-fail strategies involve the deliberate acceptance of failure in the primary operational function of drainage infrastructure while effectively mitigating the resulting implications. Tables 5.3 and 5.4 provide detailed summaries of the fail-safe and safe-to-fail approaches, characteristics, and design strategies, respectively.

Table 5.3 Comparative criteria for fail-safe and safe-to-fail techniques

	Fail-Safe	Safe-to-Fail	Source
Design Principles	Preservation of existing condition	Adaptation to changing conditions	Park et al. 2013
	Mitigation	Adaptation	Cuny 1991; Kim et al. 2017
	Risk management	Resilience	Hoang and Fenner 2015; Liao 2012
Design Objectives	Minimization of failure probability	Minimization of failure consequences	Park et al. 2013; Kim et al. 2017
	Failure prevention	Failure recovery	Seager 2008; Markolf et al. 2021
Design Focus	Component	System	Moller and Hansson 2008
	Quantitative probabilities and semi-quantitative scenarios	Possible consequences and unidentified causes	Park et al. 2013; Kim et al. 2017
Failure Impacts	Rigid/brittle	Flexible	Ahern 2011
	Rare and catastrophic	Frequent with rapid recovery	Park et al. 2013; Kim et al. 2017
Design Disciplines	Interdisciplinary	Transdisciplinary	Ahern 2013; Kim et al. 2022

Table 5.4 Design strategies and fail-safe characteristics

Design Strategies	How to Bring to a Successful End, Carry Through, or Accomplish a Desired Objective, Level, or Result
Upgrading/ Oversizing	Adding additional components to the operational system to upgrade existing drainage system infrastructure components to withstand extreme weather challenges or increasing existing drainage system and components to maximize tolerance, capacities, robustness, functionality, etc.
Redirecting/ Disconnecting	Reducing connectivity, interdependence, functionality, and interaction among the system components and between systems where those interactions already existed
Fail—Operation	Enabling the system to continue to work despite failures without major damage
Preemptive	Improving the capacity to foresee and predict changes in intense precipitation and high flood hazards and creating retrospective feedback loops between response actions to assess and develop new knowledge and adaptive strategies
Adjustments in Engineering Design Standards	Review and make necessary changes to existing design standards and protocols, which will have to change in order to encompass climate change and variability with associated uncertainties properly; adjustments in engineering design standards and changes in hazards are examined based on trend detection in observational records and projections for the future
Multi-Scale Drainage Network Connectivity/ Cohesion	Multiple elements or components of the system provide the same, similar, or backup functions by creating linkages within the system (water shade) that support and maintain function
Climate Change Adaptability	Increasing the system's capacities to change in response to new pressures and manage and account for nonstationarity design precipitation; adding additional components to the operational system to upgrade existing drainage system infrastructure components to withstand extreme weather challenges; or increasing existing drainage system and components to maximize tolerance, capacities, robustness, functionality, etc.
Adjustments in Engineering Design Standards	Review and make necessary changes to existing design standards and protocols, which will have to change in order to properly encompass climate change and variability with associated uncertainties; adjustments in engineering design standards and changes in hazards are examined based on trend detection in observational records and projections for the future

5.2.1 Safe-to-Fail Urban Drainage Infrastructure Development for the "New Normal"

Given the increasing occurrence of extreme weather events, it is essential to consider disaster preparation as a standard procedure. Resilience, the capacity of a local government to effectively adapt, respond, and restore normalcy following a catastrophic event, is a key factor in drainage infrastructure planning. Achieving resilience involves modifying both expectations and behaviors, as well as implementing appropriate design measures. Individuals should be capable of independently recovering and meeting their basic needs, such as sustenance and hydration, for the initial 72 hours following a sudden disruptive event without relying on the local government.

In an emergency, a city's likelihood of recovery significantly increases if individuals take necessary precautions and are capable of self-sufficiency. This is particularly true for individuals most susceptible to adverse circumstances, as their condition serves as a crucial measure of their ability to recover from challenges. During a catastrophic event, a municipality must effectively utilize its resources to respond and recover, aiming to preserve life, property, and commercial interests while swiftly returning to normalcy. Allocating resources to sustain human life may hinder recovery efforts due to limited resource availability.

Safe-to-fail drainage infrastructure planning is a critical component in implementing extreme weather mitigation measures. It takes into account the uncertainties inherent in forecasting and risk assessments, enabling a more efficient response to the challenges presented by climate change. Despite significant advancements in climate prediction, considerable uncertainty persists in these projections. This uncertainty primarily stems from the complex interrelations among various systems, nonlinearities in biophysical processes, the adoption of GHG-emitting technologies, and the implementation or absence of GHG-mitigation policies. This statement underscores the need for a novel approach to infrastructure development that can manage unexpected risks by building adaptable capacity while ensuring that systems do not suffer when infrastructure fails.

Most drainage infrastructure is typically designed with fail-safe principles, which involve engineering it to withstand unusual weather events and prevent potential failures. However, system failure can have significant repercussions on the environment, ecosystems, the economy, and other critical infrastructure. Fail-safe methodologies, which rely on statistical analysis of identified hazards, may prove inadequate in ensuring resilience, as they often fail to consider the uncertainties associated with nonstationary data of extreme weather. This highlights the limitations of risk-based approaches in mitigating the impacts of severe weather events and the emerging challenges associated with changing climate patterns that major urban areas are facing. Therefore, a safe-to-fail methodology, which integrates unforeseen and unpredictable risks into the decision-making process during the planning and design stages of the drainage infrastructure system, is employed. It is important to note that the concept of green infrastructure is often misunderstood as being synonymous with the idea of safe-to-fail. However, these two concepts are not necessarily mutually exclusive.

Implementing green infrastructure enhances natural processes and offers environmental, societal, and economic benefits. However, green infrastructures lacking design considerations for Type-2 failures are considered fail-safe. For instance, small-scale rain gardens can lead to ponding, localized flooding, potential health risks, and ecosystem disturbances. Green infrastructure systems like bioretention basins and green street medians are designed to account for Type-2 failures, ensuring they are safe-to-fail. These examples underscore the importance of extensive stakeholder involvement in decision making to achieve safe-to-fail drainage infrastructure development. The concept of safe-to-fail development is closely linked with the resilience of critical infrastructure. Infrastructure designed with safe-to-fail principles can offer significant advantages in terms of the financial implications of extreme events and their impact on the environment, society, and the economy relative to a fail-safe approach.

Most current drainage infrastructure systems do not strategically incorporate Type-2 failures into their design to address the potential impacts of climate change-induced extreme weather events. It is crucial to recognize that relying solely on probabilistic models and risk analyses for development decisions may not promote resilience adequately, as this approach favors the implementation of fail-safe systems that can withstand Type-1 failures effectively. The concept of safe-to-fail goes beyond fail-safe strategies and involves a multidisciplinary approach to assess the consequences of failure, specifically Type-2 failures. Policymakers should thoroughly assess the functional capacity of the drainage infrastructure, including safety thresholds, and consider the municipality's ability to address and mitigate potential risks effectively. This requires careful evaluation of various tasks across different divisions, resource allocation (human, technical, and financial), spatial condition variability, social fabric susceptibility, and service delivery capacity. Figure 5.3 illustrates how stakeholder interaction contributes to safe-to-fail development.

Integrating a safe-to-fail methodology into conventional drainage infrastructure engineering design methods presents challenges, given the uncertain and unpredictable nature of extreme weather events. Several additional challenges may hinder the effective implementation of the safe-to-fail methodology. These include decision makers' ability to understand the methodology, the performance of aging drainage infrastructure, and the need to retrofit existing infrastructure to manage excess stormwater runoff.

A key strategy for implementing a system adhering to safe-to-fail principles is to ensure the active involvement of multiple stakeholders throughout the planning and development phases. This approach ensures that various parties with vested interests in the system's success participate in the decision-making process. Incorporating diverse perspectives and expertise allows the system to be designed for resilience and adaptability to potential failures. This collaborative approach fosters ownership and accountability among stakeholders, leading to a more robust and reliable system. Engaging in a strategy that benefits decision makers when evaluating the acceptable level of failure and its financial consequences is crucial. In the context of safe-to-fail drainage infrastructure, it is noteworthy that the functions of such infrastructure can vary across cities and regions. This variation is primarily

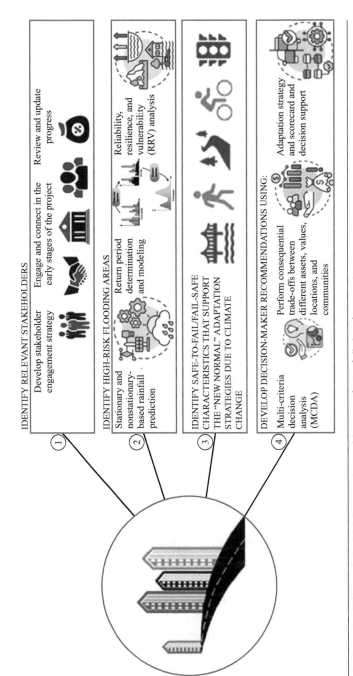

IDENTIFY RELEVANT STAKEHOLDERS

Develop stakeholder engagement strategy

Engage and connect in the early stages of the project

Review and update progress

IDENTIFY HIGH-RISK FLOODING AREAS

Stationary and nonstationary-based rainfall prediction

Return period determination and modeling

Reliability, resilience, and vulnerability (RRV) analysis

IDENTIFY SAFE-TO-FAIL/FAIL–SAFE CHARACTERISTICS THAT SUPPORT THE "NEW NORMAL" ADAPTATION STRATEGIES DUE TO CLIMATE CHANGE

DEVELOP DECISION-MAKER RECOMMENDATIONS USING:

Multi-criteria decision analysis (MCDA)

Perform consequential trade-offs between different assets, values, locations, and communities

Adaptation strategy and scorecard and decision support

Figure 5.3 How stakeholder interaction achieves safe-to-fail development

driven by the prioritization of assets, values requiring protection, and different types of failure costs.

To establish a robust and reliable drainage infrastructure system, it is paramount for each city municipality to identify critical infrastructure components and areas most susceptible to flooding. This classification is necessary to provide protection in the event of drainage infrastructure failure, which may encompass various potential scenarios such as levee breaches, bridge failures, and spillway malfunctions. This decision-making process involves evaluating the relative significance of different assets, values, locations, and individuals.

5.2.2 Safe-to-Fail Decision Criteria to Control or Minimize the Consequences of Drainage Infrastructure Failure

The frequent and unpredictable nature of extreme weather, compounded by climate change, aging drainage infrastructure, and rapid population growth and urbanization, significantly heightens the risk of large-scale flooding. Municipalities bear the responsibility of developing flood mitigation plans that prioritize the protection of the public and critical infrastructure to prevent potential harm from drainage infrastructure failure during severe weather events. The safe-to-fail methodology is a fitting strategy that allows the drainage infrastructure to withstand failures while efficiently managing or mitigating the ensuing consequences.

Conventionally, drainage infrastructure is designed to be fail-safe, meaning it is engineered to provide robust safeguards if risks are accurately anticipated and accounted for within a predetermined safety margin. However, the risks and uncertainties facing urban drainage infrastructure have significantly escalated, necessitating a reassessment of the fail-safe paradigm. The safe-to-fail principle requires planners to anticipate the possibility of failure in parts of the infrastructure system and incorporate provisions for accessing the service or purpose facilitated by the failed infrastructure. A thorough understanding of inter-system dependencies and malfunctions is essential for this task.

In the context of risk assessment for drainage systems, it is vital to consider various factors. These include the likelihood of extreme weather events or a series of events occurring, as well as the potential impact of such events on the drainage system. It is also important to consider any related or interdependent systems that may be affected. The process involves identifying potential risks arising from interdependencies between different infrastructure systems, such as energy, water, and telecommunications. Subsequently, the identified risks are prioritized based on their severity, and appropriate mitigation measures are implemented. Table 5.5 provides a severity and failure probability rating scale for assessing infrastructure failure.

Table 5.5 Severity and failure probability rating of the seriousness of infrastructure failure

Rating	Severity Category	Description
1	Slight/No Effect	Almost never/unlikely/limited damage to a minimal area of low significance.
2	Minor Risk	There are three possible outcomes related to low risks: closure, retention, or elevation to an elevated-risk group. Risks classified as low can be deemed negligible and subsequently removed from further consideration or evaluation.
3	Moderate Risk	Can be classified into two categories: high-probability/low-impact events or low-probability/high-impact events.
4	High Risk	These events are categorized as such due to two main factors: (1) they possess a significant probability of occurrence combined with a moderate level of impact, and (2) they may have a substantial impact, accompanied by at least a moderate likelihood of occurrence. In both scenarios, it is necessary to implement targeted management measures to decrease the likelihood of an event happening or to mitigate the adverse consequences of the risk.
5	Catastrophic	Imminent and immediate danger of death or permanent disability; very serious long-term environmental impairment of the ecosystem.

This scale assists stakeholders in formulating a prioritization plan for necessary actions, thereby mitigating potential hazards. When implementing protective measures, it is essential to ensure that the consequences of infrastructure malfunction are confined within the system's predetermined limits. At the onset of a project, the team must decide which failures to address, particularly if it is crucial to minimize or eliminate all failures or adopt safe-to-fail design strategies. Future designs should incorporate retrofitting solutions into existing infrastructure.

Priority scores serve as decision criteria for managing or reducing the impact of potential failures. These scores consider two key aspects: the presence of additional community benefits and the flood risk property score. The latter is a numerical assessment used to evaluate the level of risk associated with a particular property in relation to potential flooding events. However, it is essential to consider other factors when evaluating a property's suitability, such as the potential presence of supplementary benefits that the property could provide to the community.

These benefits can significantly impact the overall worth and attractiveness of a property. Therefore, it is crucial to consider these factors alongside the flood risk property score for a comprehensive assessment. Specific advancements, for instance, can enhance infrastructure, resulting in improved accessibility to various amenities and fostering economic expansion. Solely relying on the flood risk property score may not provide a comprehensive assessment of the property's overall value and potential community impact. Therefore, when assessing a property located in a flood-prone area, these supplementary factors must be considered.

When considering flood mitigation, it is crucial to evaluate multiple factors to determine the priority scores, which have been formulated to encompass a broad spectrum of factors. Table 5.6 provides valuable insights into developing a safe-to-fail system for effectively managing and mitigating the potential ramifications associated with drainage infrastructure failure. It presents the decision criteria, priority, and mitigation technique scores, which are integral elements of the design process. By integrating these scores, engineers and designers

Table 5.6 Example decision criteria priority and mitigation technique scores to be incorporated in a safe-to-fail design

Decision Criteria Priority	Scores	Mitigation Techniques	Scores
Life and human safety		Property acquisition, structure demolition, rebuild, relocation, and structure elevation	
Cost-effectiveness		Infrastructure upgrades, drainage systems, sewers, and culverts	
Proximity to other mitigation projects		Residential programs; sewer backflow and downspout disconnection and building code changes	
Property added to flood zone		Structural and nonstructural flood control and coastal defense, land-use change, levee/floodwall protection for multiple structures, and dry/wet floodproofing of structures	
Repetitive loss structure		Zoning, acquisition and land-use regulation, and socioeconomic incentives	
Property located in close proximity to publicly owned land		Audible flood warning system for individual property, automated flood notifications, and public education	
Property situated in close proximity to various planned mitigation projects		Implementation of flood-resistant design and construction practices and incorporation or enhancement of *freeboard* criteria within flood damage ordinances	
Property situated on a designated stormwater drainage or storm sewer easement		Creation of engineering standards for the integration of drainage systems in recently developed regions, emphasis on the encouragement of low-impact development techniques and green infrastructure practices	
Property intersects with water quality buffer		Integration of protocols for monitoring and documenting peak water levels subsequent to a flood event within emergency response blueprints	

Continued

Table 5.6   *continued*

Decision Criteria Priority	Scores	Mitigation Techniques	Scores
Property located in an environmental focus area		Redirection of excess stormwater runoff resulting from heavy rainfall or flooding incidents toward specifically designated locations, such as detention and retention structures, large-scale green infrastructures, or other retrofitted stormwater management structures. Enhancement of the ecological integrity and functionality of water bodies to mitigate the environmental consequences associated with urban development and infrastructure initiatives	
Historic preservation and cultural asset protection		Formulation of an integrated assessment framework for evaluating the effectiveness of various adaptation strategies, including risk analysis and modeling, geographic information system (GIS) application, and transportation vulnerability networks analysis. This integrated approach enables the evaluation of the potential impact of different factors and facilitates well-informed decisions regarding adaptation measures	

The scoring system should assign numerical values to different issues by considering factors like urgency, impact, complexity, risk, or dependency. It can utilize various scales, including predefined ranges 1 to 10, or even custom formulas like (urgency × impact)/(complexity × risk). The scoring system should be characterized by transparency, objectivity, and consistency, with regular updates occurring as issues arise. Utilize the scores to effectively rank the issues or set a priority threshold, for example, categorizing only issues with a score above 6 as *high priority*.

can proficiently evaluate and mitigate potential risks linked to drainage infrastructure failure. This systematic approach ensures the implementation of suitable measures to minimize the consequences of these failures (see Table 5.7), ultimately improving the system's overall safety and dependability.

Table 5.7 Cost categories resulting from the severity and failure probability rating and associated impact

Cost Category	Associated Impact
Direct tangible costs	Costs associated with infrastructure damage, particularly in relation to emergency services and disaster assistance—infrastructure damage refers to the destruction or impairment of essential physical structures, such as roads, bridges, buildings, and utilities, crucial for societal functioning—this occurs when a disaster strikes, either natural or man-made
Indirect costs	Expenses related to reconstruction and recovery, along with costs associated with planning and executing risk prevention measures—these factors are vital in reducing the impact of disasters and safeguarding community welfare and security; in the event of a disaster, the primary task is rebuilding and recovering, which entails restoring or replacing impaired infrastructure, residences, and public amenities, and reinstating vital utilities such as electricity, water, and transportation networks

Direct social loss	• Loss of life
	• Various ways individuals are impacted by unfortunate circumstances such as going missing, being displaced, becoming homeless, and experiencing loss of livelihood—these situations can significantly impact communities
	• Property loss
	   i. *Real property*: land-related assets include the land and any permanent constructions, natural elements growing on the ground, foundations, subterranean pipes, and other land-based components
	   ii. *Personal property*: any tangible or intangible item in a wide range that falls under different classifications

Losses incurred as a result of business interruption	Losses due to business interruption encompass various factors, including revenue decline, which directly impacts the financial performance of a business; the absence of basic public services such as transit, internet access, water, electricity, and gas can exacerbate these losses—these services are vital for facilitating efficient business operations, and their absence can result in substantial adverse consequences
Intangible costs	These include health impacts, environmental losses, cultural heritage losses, and psychological stress

A criteria rating matrix, also known as a grid analysis, is a systematic method used to evaluate each solution option or mitigation strategy. This method is based on the relevant criteria required for decision making regarding solution implementation. It enables an objective evaluation of all available alternatives. The matrix is organized with options as rows and evaluation criteria as columns, simplifying the analysis process. Each alternative is evaluated based on its ability to meet each criterion.

5.3 SCENARIO-BASED ROBUST DECISION MAKING (RDM)

RDM is a scenario-based methodology that aids decision makers in navigating uncertainties and potential risks. It is widely employed across various disciplines, enabling the formulation of flexible and informed decisions amidst substantial uncertainty. RDM involves analyzing multiple potential future outcomes and assessing their viability, advantages, and disadvantages to devise effective decision-making strategies. The approach prioritizes robustness over optimality, aiming to guide policymakers in making well-informed, long-term decisions. It evaluates potential short-term strategies across a wide array of possible future scenarios, with the primary goal of helping policymakers forecast or mitigate the adverse effects of unforeseen events. These events result from the interaction of external uncertainties, or exogenous factors, beyond their control, with measures within their jurisdiction.

"Deep uncertainty" refers to decision-making scenarios lacking probabilistic information, making decision making challenging. RDM is applicable in situations where conventional risk information, such as defined probability distributions, is unavailable or where there is disagreement on the models or methods to use for evaluating the desirability of different outcomes. RDM aims to support decision makers in making resilient decisions under current circumstances, despite incomplete and uncertain future knowledge. *Robust optimization*, a methodology that considers uncertain parameters typically viewed as static in deterministic approaches, aids in identifying the optimal design of a drainage system within budget constraints.

RDM is formally applied using a modeling interface guided by both human and computer input, combining qualitative and quantitative information. This combination offers superior analytical capabilities relative to traditional decision support tools. The use of historical pattern reorganizations and data-driven analysis is crucial in evaluating the performance of existing assets in large-scale urban watershed drainage networks concerning proposed development. This process employs incremental and dynamic data analysis techniques across various locations to identify potential vulnerabilities or weaknesses associated with the proposed development and existing assets. The approach is formally applied using data mining algorithms that analyze vulnerability and response options.

To successfully implement RDM, a series of systematic procedures must be executed. As depicted in Figure 5.4, the fundamental process of RDM begins with problem structuring, not the immediate identification of significant uncertainties or threats. This approach allows for a comprehensive understanding of the problem before addressing potential risks or uncertainties. By initiating problem structuring, practitioners can establish a solid foundation for analysis, ensuring all relevant factors are considered. This systematic approach enhances the effectiveness and accuracy of subsequent steps in the RDM analysis process, which includes the generation of as many future scenarios as possible. This is achieved by using computer modeling and data aggregation analysis to efficiently generate alternative strategies for consideration and prioritization. A comprehensive analysis should be conducted to evaluate the uncertainties and factors associated with each strategy across a wide range of possible futures to determine the course of action. Uncertainty values are determined for each variable within a specified range based on stakeholder input or alternative methodologies. Subsequently, a computer algorithm generates a wide range of prospective future scenarios to evaluate the efficacy of each strategy. The statistical derivation of the most significant uncertainty

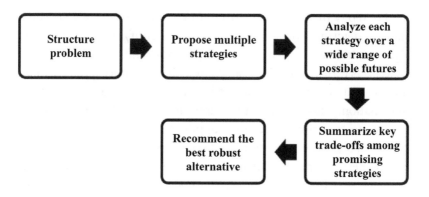

Figure 5.4   Basic robust decision-making process

parameter combinations for strategy selection is computed, followed by the development of a summary outlining the primary trade-offs among the most favorable strategies.

Performance measures are employed to assess predetermined, favorable outcomes when scrutinizing strategies across diverse scenario futures. The primary aim of RDM is to pinpoint a robust strategy capable of yielding favorable outcomes across a broad spectrum of circumstances. If an effective and robust strategy is not identified, an iterative process of strategy reformulation begins, involving stakeholders. RDM encompasses the systematic evaluation of various strategies against computer-simulated future scenarios (refer to Figure 5.5). This process considers external factors beyond the decision maker's control, as well as policies or options within their purview. Functional relationships are established between these factors, and each strategy's performance is assessed using quantitative measures. The initial approach involves the use of statistical or data mining algorithms.

5.3.1 Application of Scenario-Based Robust Decision Making to the "New Normal"

RDM and adaptive management (AM) are closely related and share several key attributes. Both approaches recognize and embrace uncertainty and are widely employed across various industries to enhance

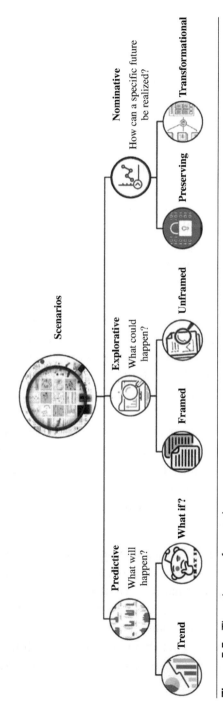

Figure 5.5   Three types of scenarios

decision-making processes. AM is a systematic approach that involves continuous learning from decision outcomes and subsequent adjustments.

Both RDM and AM are valuable tools for addressing complex and unpredictable problems. However, RDM is often recommended as a suitable approach for adaptation. It aids in identifying strategies, policies, or options that demonstrate robustness, meaning they are satisfactory for their intended purpose. This approach provides an alternative to traditional cost-benefit analysis (CBA), which is typically used to identify the most efficient mitigation options based on monetary considerations. RDM is particularly useful for assessing the impact of extreme weather events on drainage infrastructure systems using a predict-then-optimize framework.

RDM is gaining popularity as a decision analysis approach in situations of significant uncertainty. Various metrics have been proposed to measure the efficiency of research data management. The application of resilience parameters can generate distinct scores for decision-making options. These potential contributing factors in the computation of robustness metrics are presented for clarification, aiming to provide a clear understanding of how robustness metrics function, their appropriate usage, and the factors that may contribute to any inconsistencies that arise.

The use of different metrics to evaluate the relative robustness score of decision-making options can result in either consensus or divergence among the results. These metrics measure the consistency of decision alternative rankings across robustness metrics. Several robustness metrics are used to assess a system's efficacy in handling substantial uncertainty:

1. Quantifying statistical properties beyond the mean, such as variance and skewness, can provide insights into the variability of expected performance levels across various scenarios.
2. Expected value metrics estimate the anticipated average value for given scenarios and performance levels.
3. Optimizing and Satisfying metrics establish the range of scenarios that meet a satisfactory performance threshold.

4. Regret-based metrics quantify the discrepancy between the performance of a selected option under a certain condition and the optimal performance that could have been achieved under the same condition.

These metrics evaluate the system's performance across various future scenarios. The lack of a well-defined standard for measuring an option's robustness presents a challenge in determining an option's overall resilience and comparing it with other options. In decision making, it is common for individuals to experience confusion when comparing the strength values and rankings of various decision alternatives. This confusion often arises when different robustness metrics have been used to evaluate these alternatives. This issue can be addressed via a cohesive framework that calculates a variety of robust metrics:

1. A standardized framework for computing various robustness metrics is beneficial for comparing robustness values obtained from different metrics in an unbiased manner. This framework allows for the computation of various robustness metrics, enhancing the efficiency and reliability of the comparison process. By using this standardized approach, decision makers can ensure that their robustness evaluations are consistent and easily comparable across different contexts.

2. A systematic classification system for robustness metrics can create a methodical categorization system. This system aids in selecting the most suitable metric for a given scenario and provides decision makers with guidance on choosing an appropriate robustness metric.

3. A theoretical framework can probe the factors influencing the outcomes of different decision-making processes based on various robustness metrics. The framework can also assess the stability or robustness of alternative rankings when different robustness metrics are employed. Its primary purpose is to provide supplementary guidance to decision makers.

4. The conceptual framework is tested using different decision contexts, targets, scenario categories, and alternative choices to determine its efficacy and applicability.

5.3.2 Calculating Scenario-Based Robustness Metrics

The calculation of robustness metrics, a set of procedures used to evaluate a system's resilience to variations and disturbances, involves analyzing the system's performance under different conditions and assessing its ability to withstand changes in input or environmental factors. The methodology for calculating these metrics can vary, depending on the system under evaluation and the intended use of the metrics. The process typically includes defining relevant performance criteria, selecting appropriate test scenarios, and conducting experiments to collect data on the system's behavior. Subsequently, statistical methods are used to derive numerical measures of the system's resilience.

Robustness metrics are extensively employed to evaluate the resilience of various decision-making options, including local government initiatives, municipal ordinances, design standards, guidelines, regulations, policies, innovative solutions, and operation and maintenance management plans, among others. When calculating any robustness metric, three essential elements must be specified for accurate determination of the desired metric:

1. The decision alternatives under consideration for robustness determination. These alternatives must be carefully evaluated to ascertain their robustness level.
2. The performance metric of interest, which could be influenced by factors such as cost or reliability. Focusing on these metrics allows for a better understanding of the alternatives' performance in different scenarios.
3. The plausible future scenarios used to evaluate the alternatives' performance.

Considering a range of possible scenarios is crucial for assessing the robustness of decision alternatives. By taking into account these three elements, we can gain valuable insights into the robustness of different alternatives and make more informed decisions. Figure 5.6 illustrates the three components that constitute robustness.

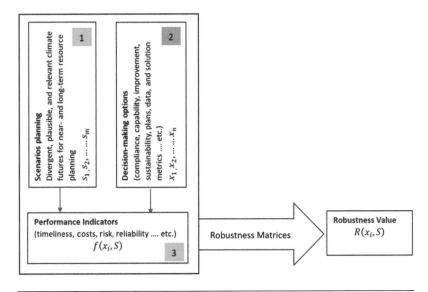

Figure 5.6 Three components of scenario-based robustness

The assessment of robustness is a crucial decision-making step. It involves thoroughly evaluating a particular decision alternative x_i, across a predetermined range of future scenarios $S = |\; s_1, s_2, s_3 \ldots \ldots \ldots s_n \;|$, using a designated performance metric, $f(x_i, s_i)$. The process allows for evaluating the resilience and effectiveness of the alternative under various conditions. It involves transforming $f(x_i, S)$ (the performance decision alternatives) into corresponding robustness values, denoted as $R(x_i, S)$. This transformation helps assess the alternative's ability to withstand uncertainties and variations in different scenarios. By considering multiple potential outcomes, it is possible to enhance our comprehension of the potential risks and rewards associated with the decision under consideration.

The transformation of $f(x_i, S)$ into $R(x_i, S)$ can be achieved through various robustness metrics, each with its unique approach. However, a cohesive framework can be established by breaking down the overall transformation into three distinct stages:

1. *Performance Value Transformation* (T_1): This refers to the process of converting raw performance values into a more meaningful and standardized form. By doing so, we can compare and evaluate different performance measures more effectively.

2. *Scenario Subset Selection* (T_2): This involves carefully selecting a smaller, more manageable subset of scenarios from a larger set. To accurately depict a particular situation or problem, it is necessary to ensure correct representation. By employing a meticulous approach in choosing a specific set of scenarios, we can proficiently examine and comprehend the issue at hand without becoming overwhelmed or sidetracked.

3. *Robustness Metrics Calculation* (T_3): This refers to the calculation of robustness metrics to determine the quantitative measures that evaluate the resilience and reliability of a system or process.

These stages serve as the fundamental building blocks for the framework, offering a systematic approach to the process of transformation. Dividing the transformation into different stages allows us to analyze and manage each step separately, ensuring smooth and efficient progress toward achieving the desired outcome. Figure 5.7 illustrates the common scenario-based robustness assessment metrics framework.

Metrics that depend on absolute performance values, such as cost and reliability, remain unchanged under an identity transformation, represented as T_1. Various metrics can be employed to assess a system's resilience, transforming the system's absolute performance values into alternative ones. These transformed values are used to quantify the extent of regret, thereby assessing the implications of choosing one decision alternative over another. The term *regret value* denotes the potential for a different choice to have yielded more favorable outcomes in a specific future scenario. It is crucial to note that decision alternatives can also serve as indicators for evaluating whether the chosen option will yield satisfactory performance and whether the necessary constraints have been met.

The selection of a subset of scenarios T_2, known as the scenario subset selection transformation, is critical in calculating the robustness

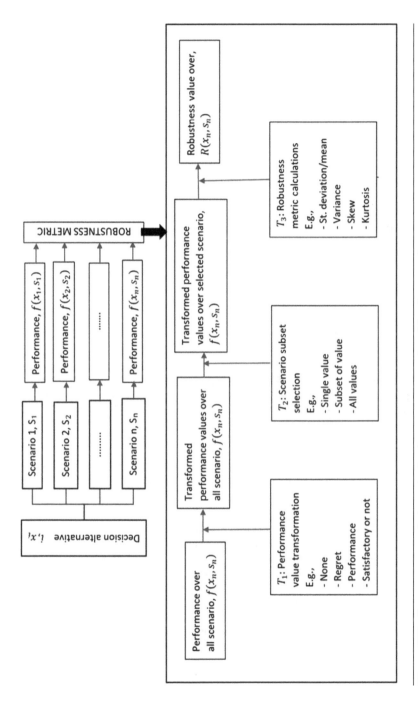

Figure 5.7 Common scenario-based robustness assessment metrics framework

metric T_3. The T_2 transformation involves selecting specific values for use in the calculation, akin to selecting a subset of scenarios for performance evaluation, represented as $|f'(x_i, s') * f'(x_i, S)|$. The consideration of extreme scenarios during the calculation of a robustness metric indicates a heightened focus on risk prevention. Conversely, excluding these extreme scenarios implies a lower emphasis on risk avoidance. Table 5.8 presents commonly used scenario-based robustness metrics.

Table 5.8   Commonly used scenario-based robustness metrics

Spread	The spread evaluates the performance range, calculated as the difference between maximum and minimum values. It does not consider average or median performance metrics, implying that a consistently low-performing structure could still be deemed resilient. The spread calculation is performed for every configuration, identifying the performance delta between optimal and suboptimal results across all conceivable scenarios. The configuration with the smallest difference between its highest and lowest values is considered the most resilient.
Deviation	Deviation refers to the divergence or variation of a data point or set of data points from a central value or average. It provides information about the spread or dispersion of the data, aiding in understanding the overall pattern or distribution. This metric facilitates the computation of both worst-case and best-case values across various configurations and scenarios. The worst-case value pertains to the minimum achievable outcome within a specific configuration or scenario, while the best-case value denotes the maximum attainable result across all configurations and scenarios.
Minimax Regret	This is a computational approach to decision making that seeks to minimize the expected maximum regret across various alternatives. It involves evaluating the potential outcomes of each decision and choosing the alternative that would yield the lowest level of subsequent remorse. By incorporating an analysis of the potential negative outcomes associated with each choice, this strategy enables decision makers to make more informed and cautious decisions.

Continued

Table 5.8 *continued*

Maximin	The maximin criterion is a commonly utilized approach in decision making. This concept of regret involves evaluating the discrepancy between the best possible choice and the actual decision made in a given situation. This metric evaluates different configurations by identifying the scenario that produces the lowest performance for every setup and then selecting the combination that demonstrates the highest outcomes in that worst-case scenario.
Percentile-Based Metrics	These metrics provide a way to assess the likelihood of different outcomes. They can be particularly helpful when decision makers want to understand the potential range of results and make decisions based on specific percentiles. By integrating these additional metrics, decision makers can acquire a more holistic understanding of the available alternatives.
Taguchi Metric	This metric, formulated based on the principles and methodologies developed by renowned Japanese engineer and statistician Genichi Taguchi, enables decision makers to effectively measure and analyze performance. This methodology involves the application of statistical techniques, such as computing the mean and standard deviation, to analyze and quantify the impact of various factors or variables on the specified process.
Professor Starr's Methodology	Professor Starr's domain criterion, a concept commonly used in various fields, refers to a specific set of conditions or requirements that must be met within a particular domain. This metric involves evaluating the performance of a given entity relative to a decision-maker-determined threshold value. Defining a performance acceptance threshold is a crucial step for the decision maker. The process involves converting the performance matrix into a Boolean matrix, also known as a logical matrix. The user calculates the mean acceptance ratio for each configuration.
Global Maximum (Maximax)	This metric identifies the configuration that performs best under optimal conditions. Although not a measure of robustness, it can be useful in interpreting and facilitating discussion of the outcomes.
The Laplace Principle of Insufficient Reasoning	It is important to note that this metric operates under the assumption of equal probability for all scenarios. This tool assesses the appropriateness of using average performance as a metric for evaluation.

During the third phase of the transformation process, the actual robustness metric of the system is calculated using the transformed performance values from T_1 and applying them to the selected scenarios in T_2. The result is a single robustness value, $R(x_i, S)$, derived from the transformed function $f'(x_i, s')$. If a single scenario is chosen in T_2, the resulting performance value is automatically considered the robustness value due to the presence of only one transformed performance value, equivalent to an identity transform. In situations with multiple scenarios and transformed performance values, statistical moments must be calculated to consolidate them into a single value.

5.3.3 Scenario-Based Robustness Metrics Framework

The appropriate use of various robustness metrics is a complex task that requires careful consideration and analysis. The selection of the suitable metric is crucial in ensuring the reliability and consistency of the decision option. As the team deliberates on the optimal approach for their project, decision makers must understand the importance of examining the taxonomy of different robustness metrics. The objective is to comprehend each metric thoroughly along with its potential applications. Armed with this knowledge, they can make an informed decision that will ultimately benefit the project. Scenario-based robustness metrics can be classified according to different transformation functions. The following is a typical scenario-based robustness transformation function:

1. $f(T_1) = f(x_n, s_n)$: This transformation function (T_1) emphasizes performance value across all scenarios.
2. $f(T_2) = f(x_n, s_n)$: This transformation function (T_2) involves the identification and extraction of a subset of scenarios from a larger pool.
3. $f(P_1) = p(x_n, s_n)$: This transformation function (P_n) simulates each configuration's performance for every scenario, resulting in the creation of a performance matrix. At this stage, the currently used hydrology and hydraulic simulation model is

utilized. Performance can be measured using various metrics, including the frequency and intensity of heavy precipitation events, flooding, drainage structure failure, cost, and other relevant factors.

4. $f(T_3) = f(x_n, s_n)$: This transformation function (T_3) involves the calculation of a robustness value. Computation is the assessment of the project's overall success.

The metrics consider the overall effectiveness of each decision alternative in a specific scenario, including factors such as cost and reliability. The values of $f(x_i, S)$ comprise performance values. The robust decision alternatives maximize the system's performance across all scenarios (see Figure 5.7).

5.4 ROBUST SOLUTION ANALYSIS AND INTERPRETATION

After simulating the performance of each configuration under every scenario, the next step is a robustness analysis. Different industries use various metrics with degrees of risk aversion to aid decision making. These robustness metrics, also known as robustness indicators, assessment approaches, or methods, are outlined in Table 5.8. In decision analysis, minimax regret and spread are two commonly used techniques. The maximin criterion, which selects the option with the highest degree of pessimism, is a valuable tool for identifying the most resilient choice.

Regarding Laplace's principle of insufficient reasoning, it is crucial to determine the appropriateness of using average performance as an evaluation metric. In the context of performance evaluation against a predetermined threshold, the domain criterion proposed by Professor Starr is used. This criterion serves as a standard for making decisions or judgments and provides a framework for evaluating and comparing different options.

In addition to standard metrics, other tools, such as the Hurwicz optimism-pessimism rule, developed by economist Leonid Hurwicz, can be used in decision making. This rule, which considers the decision maker's level of optimism or pessimism, allows for a nuanced analysis by considering the best and worst possible outcomes. Another approach is to use percentile-based metrics, which assess the likelihood of different outcomes. These metrics are useful when decision makers want to understand the potential range of results and make decisions based on specific percentiles. By integrating these additional metrics, decision makers can gain a more comprehensive understanding of the available alternatives.

The robustness analysis quality depends on the comprehensiveness and accuracy of the scenario definition. Consideration of system performance should include an exploration of extreme scenarios and system response. However, scenarios should be selected based on predictions of potential future developments. Once the most robust solution is identified, it must be carefully interpreted. Interpretation should be entrusted solely to individuals with expertise in the relevant scenarios due to the potential for errors in interpreting the results obtained from performance distribution and probabilistic analysis. Performance distribution, the statistical representation of a system's performance, enables us to understand and evaluate the effectiveness of different systems or processes.

Performance distribution pertains to the distribution of performance among individuals or entities. It involves probing the variation and spread of performance levels across a population. This analysis can yield valuable insights into the overall performance landscape and help identify key factors. Probabilistic analysis is a technique used in various fields to understand the underlying probabilities and uncertainties associated with a system or phenomenon. It allows for assessing the likelihood of different outcomes and making informed decisions per the given information. This approach is beneficial when dealing with complex systems or situations with inherent variability and randomness. Through probabilistic analysis, we can gain valuable insights into the system's behavior and characteristics.

Decision makers must always be prepared for future developments. This is where robustness analysis proves useful. By conducting a robustness analysis, decision makers can identify the most resilient choices that can withstand potential future changes. This analysis is especially useful when making decisions with long-term impacts. By identifying the most robust choices, decision makers can ensure their decisions remain effective and relevant, even in the face of uncertainty. When used alongside sensitivity analysis, robustness analysis becomes a powerful tool for gaining a deeper understanding of potential future performance.

5.5 CHAPTER SUMMARY

Climate change has led to a significant acceleration in sea-level rise, the emergence of extreme weather events—often referred to as "New Normal" events—and ecosystem disturbances. Thus, urban areas with extensive infrastructure are exposed to various emerging risks. In line with established engineering standards for risk assessment, it is essential for professionals to recognize the increasing impact of unpredictable weather events. These events have caused considerable disruptions in infrastructure, leading to severe and catastrophic consequences. These failures result in loss of life, property damage, disruption of public services and critical infrastructure, business operation interruptions, health hazards, environmental damage, and negative impacts on regional economies. Infrastructure development should include a systematic examination and evaluation of potential risks, followed by the implementation of appropriate measures to enhance reliability and mitigate potential disruptions. Climate models have been instrumental in predicting the potential effects of severe weather phenomena on infrastructure. Factors influencing weather forecasts include climate dynamics, human systems, and natural resource usage.

Frequency analysis has become a widely used technique for determining the optimal capacity of urban drainage infrastructure and assessing the associated flood risks. The field of hydrology has traditionally relied on the practical assumption of stationarity. However, in recent decades, the effects of climate change and land-use changes have led to the emergence of time-series trends, indicating the presence of nonstationary effects. The application of nonstationary analysis significantly improves the accuracy of urban drainage hydrology and hydraulics analysis, facilitating the development of flexible drainage systems that can adapt to climate and land-use changes, ensuring optimal functionality under different conditions. There is a need to enhance existing drainage design guidelines by incorporating a comprehensive methodology for nonstationary frequency analysis.

Historical and current infrastructure development practices often lack adequate integration of risk assessment and mitigation measures to address potential failure scenarios throughout the entire development life cycle. Current infrastructure primarily prioritizes service delivery optimization within financial and safety regulation constraints. Current development methodologies are insufficient due to their failure to integrate robust infrastructure with low failure rates, long lifespans, and the capacity to accommodate unforeseen risks, ultimately leading to failures.

Implementing a fail-safe approach that prioritizes immediate reliability and risk reduction could inadvertently increase future potential harm. As infrastructure scales up and becomes more durable, the potential consequences of its failure also increase significantly. While it might seem feasible to incorporate failure outcomes into risk analyses, even the most advanced models cannot accurately predict future nonstationarities, such as severe weather events, demographic changes, urban growth, and policy alterations. Moreover, the results generated by these models may not provide a comprehensive understanding of the system's state when external forces exceed the system's operational capacity. While climate model forecasts can enhance the robustness or safety of local elements, current computational simulations do not fully consider the potential consequences of system failure. Therefore, engineers and decision makers who are involved in various stages of infrastructure development must recognize the potential for unforeseen failures not accounted for in current models.

Oversizing—a technique developed specifically to enhance resilience—has been widely used as the primary approach to mitigate failure in fail-safe infrastructure. Design standards play a crucial role in mitigating the risk of oversizing a project by establishing precise requirements that ensure the project's robustness, functionality, safety, longevity, and practicality. Using design standards that account for changing stress factors on systems, such as the increasing number and severity of storms, changes in water sources, groundwater depletion, extreme heat, and environmental loads, can make infrastructure more resilient to future weather conditions. However, the oversizing approach, while reliable, does not consider potential future climate variations and is therefore not considered an efficient strategy in the context of nonstationary climate conditions characterized by high uncertainties. This discrepancy arises from the potential for a significant divergence between the analytical forecasts generated by climate models and the range of design criteria.

Developing resilient infrastructure is crucial for implementing climate change adaptation strategies that account for inherent uncertainties in climate models and risk analysis. Despite significant advancements in climate prediction, these projections still exhibit considerable uncertainty. This uncertainty stems from multiple factors, including complex system interconnections, nonlinear behavior of biophysical processes, extensive use of GHG-emitting technologies, and the uneven adoption of GHG reduction policies. This underscores the need for an innovative approach to infrastructure development that effectively addresses unexpected risks by building adaptable capacity and ensuring urban system robustness in case of failure.

In traditional approaches, infrastructure is meticulously designed to incorporate fail-safe mechanisms. These measures aim to mitigate the impact of infrequent meteorological phenomena and ensure uninterrupted infrastructure operation. However, the consequences of failure, which could include human casualties, significant economic impact, and substantial infrastructure damage, are of considerable magnitude. Fail-safe approaches based on risk assessment often rely on statistical analyses of identified risks but frequently fail to consider the uncertainties associated with climate change. Consequently, they lack the necessary capabilities to ensure resilience.

The safe-to-fail concept is highly significant in preparing for a future where climate change will substantially influence the management of uncertain and unidentified risks during urban drainage infrastructure development. Safe-to-fail development is closely linked to the resilience of technological, ecological, and social systems. The advancement of existing infrastructure will significantly impact future costs associated with climate change, including its social, environmental, and economic consequences. Municipalities actively participate in infrastructure development to effectively mitigate and adapt to the diverse challenges posed by climate change. Relying solely on climate models (both stationary and nonstationary) and risk analyses to guide development decisions may not effectively support resilience advancement. This methodology enables the creation of fail-safe systems specifically designed to withstand Type-1 failures. To make something safe-to-fail, practical engineering methods must be combined with a multidisciplinary approach that considers potential failure scenarios and their consequences.

Safe-to-fail development requires a comprehensive examination of the biophysical capabilities of infrastructure, including safety thresholds, as well as the social capacities necessary for effective risk management. Social capacities refer to a society's collective ability to efficiently address diverse challenges and meet its constituents' needs. These capacities can be understood by analyzing four fundamental dimensions: institutional capacity, spatial variability, social vulnerability, and serviceability. There is often a lack of precision in distinguishing between green infrastructure and safe-to-fail, despite their distinct conceptual differences. Green infrastructure is a strategic approach that optimizes natural systems to enhance their functionality while simultaneously delivering a range of environmental, societal, and economic benefits. However, deploying green infrastructure without considering Type-2 failures lacks the integration of fail-safe mechanisms. A common phenomenon is ponding in smaller-capacity rain gardens, which can potentially flood the surrounding area. This scenario can negatively affect the overall welfare of the population and disrupt the ecological system.

Green infrastructure systems, such as bioretention infiltration-based best management practices (BMPs), are specifically engineered to mitigate Type-2 failures. This enhances their safety and durability in the event they occur. This example underscores the importance of prioritizing decisions by extensively involving stakeholders to implement a safe-to-fail development approach. However, implementing safe-to-fail development faces significant challenges due to the uncertain nature of the risks and performance associated with long-lasting infrastructure. Moreover, incorporating strategies aimed at achieving *safe failure* can hinder the use of practical engineering methodologies. Adapting to new stress factors after construction and managing costs arising from decisions made during development can be challenging, especially when dealing with rigid infrastructure systems.

Successful infrastructure development, which allows for a certain level of variation, requires a comprehensive understanding and a clearly defined decision-making framework. The development process should consider multiple factors, such as the precise geographical context, preexisting infrastructure services, social vulnerability, various types of failure costs, and the capacity for institutional adaptation. One strategy for establishing a safe-to-fail environment involves using multistakeholder engagement to help decision makers determine the acceptable threshold for failure and its associated costs. The functionalities of safe-to-fail infrastructure can vary across different cities and regions due to differences in asset prioritization, value protection, and the ability to handle different types of failure costs. Ideally, it is possible to develop a resilient infrastructure system for each urban region by carefully identifying the entities or components that need protection in the event of infrastructure failures. This decision-making process involves assessing the various assets, values, locations, and individuals involved and making appropriate trade-offs.

5.6 CHAPTER PROBLEMS

1. The municipality of "New Normal" experiences three distinct types of flooding: precipitation-induced, tidal, and coastal. Each type, resulting from intense rainfall, tidal fluctuations, and storm surges, respectively, disrupts the stormwater conveyance system. In the initial phase of the city's flood risk mitigation strategy, a comprehensive assessment identifies areas needing

improvement. Upgrades are necessitated by the current drainage infrastructure system's inadequate capacity, which includes bridges, culverts, and storm sewer pipes. Certain areas, due to their low elevation, are naturally prone to flooding and often experience such during extreme weather events. The municipality also addresses flooding complaints, locations requiring intensive maintenance, and existing drainage infrastructure capacity and maintenance information. Budgetary costs have been meticulously calculated and analyzed to allocate city funds efficiently, aiming to mitigate the impact of extreme weather flooding. The city has also identified priority project areas for each planning district to concentrate efforts on the most vulnerable areas, thereby maximizing resource effectiveness and minimizing flood risks.

"New Normal" has contracted a consulting engineering firm to develop a comprehensive flood mitigation master plan. This plan will prioritize regions within the municipality identified as having a higher susceptibility to flooding. The engineers will utilize existing data and assess the current condition of the stormwater drainage infrastructure to formulate this plan. The project will be financed using the budget allocated for the Capital Improvement Program in the 2025 fiscal year.

The consulting engineering firm is expected to submit its proposed framework of scenario-based robustness metrics for an optimal, reliable design of storm drainage systems for city approval. The city has established four objective functions that must be met:

a. Minimize construction costs
b. Determine the appropriate pipe diameter upgrades
c. Minimize the occurrence of floods and identify the most effective BMPs
d. Maximize infiltration to recharge the groundwater for the underlined soil with a 25 mm/hr infiltration rate

The problem is formulated as follows:

The contracted consulting firm is tasked with developing objective functions that will minimize identified problems and maximize the effectiveness of the city's flood risk mitigation strategy. The decision function will be determined by the variables of pipe diameters and the type and proportion of BMPs to be implemented in the targeted catchment areas. Uncertainty in these functions arises from three primary factors: temporal variations in stationary rainfall patterns, degradation of pipe roughness over time (n-value), and alterations in land use within each subcatchment impacting its permeability:

$$\min{}_{D,B} flooding \; [D, B, H, n, R_c] \; \dotfill \; 1$$

$$\min{}_{D,B} [D, B] = min_{D,B}(C_D * L) + (C_B * B) \dotfill 2$$

$$\min{}_{D,B} runoff \,(Q) \; [D, B, H, n, R_c] \; \dotfill \; 3$$

$$C_D, D, L, n \in R^p \; \dotfill \; 4$$

$$Cost, Flooding, Runoff \,(Q) \in R$$

$$H \in R^t$$

$$R_c \in R^s$$

$$C_B, B \in R^a$$

Let D and B denote the decision vector variables representing the pipe diameters and the proportion of BMPs implemented in the basin, respectively. City employees have approved hydrology and hydraulics modeling software, including HEC-HMS/RAS/GeoRAS (ACoE), PRMS, SWMM/XPSWMM (EPA), INFOWORKS, StormCAD (Bentley), and HydroCAD. These tools calculate parameters such as precipitation, stormwater runoff discharge (Q), volume (V), and flood (f). C represents the monetary value associated with the construction of the storm drainage system. The hyetograph vector is denoted by H, indicating the hyetograph, with t representing the number of discrete time slots on the vector. R_c and n represent the imperviousness of the subcatchment and the degradation of the pipe roughness vector over time, respectively. (C_D) is a mathematical vector quantifying the cost per unit length of a pipe, considering the various diameters (D) of the pipes. The vector (L) represents the length of each individual pipe within the network. The cost vector for the BMPs, denoted as (C_B), depends on the specific BMPs (B) used. P, a, and s represent the quantities of pipes, areas utilizing BMPs, and subcatchments within the basin, respectively.

SELECTED SOURCES AND REFERENCES

Abramson, D. M. and I. Redlener. 2012. "Hurricane Sandy: Lessons Learned, Again." *Disaster Medicine and Public Health Preparedness* 6, no. 4: 328–329.

Ahern, J. 2011. "From Fail-Safe to Safe-to-Fail: Sustainability and Resilience in the New Urban World." *Landscape and Urban Planning* 100, no. 4: 341–343.

Ahern, J. 2013. "Urban Landscape Sustainability and Resilience: The Promise and Challenges of Integrating Ecology with Urban Planning and Design." *Landscape Ecology*, 28, 1203–1212.

Beh, E. H. Y., F. Zheng, G. C. Dandy, H. R. Maier, and Z. Kapelan. 2017. "Robust Optimization of Water Infrastructure Planning Under Deep Uncertainty Using Metamodels." *Environmental Modelling & Software* 93: 92–105.

Beh, E. H. Y., H. R. Maier, and G. C. Dandy. 2015. "Adaptive, Multiobjective Optimal Sequencing Approach for Urban Water Supply Augmentation Under Deep Uncertainty." *Water Resources Research* 51, no. 3: 1529–1551.

Ben-Haim, Yakov. 2004. "Uncertainty, Probability and Information-Gaps." *Reliability Engineering and System Safety* 85, no. 1: 249–266.

Bennis, W. M., D. L. Medin, and D. M. Bartels. 2010. "The Costs and Benefits of Calculation and Moral Rules." *Perspectives on Psychological Science* 5, no. 2: 187–202.

Ben-Tal, A., L. El Ghaoui, and A. Nemirovski. 2009. *Robust Optimization*. Princeton and Oxford: Princeton University Press.

Bertsimas, D. and M. Sim. 2004. "The Price of Robustness." *Operations Research* 52, no. 1: 35–53.

Bilal, K., M. Manzano, S. U. Khan, E. Calle, K. Li, and A.Y. Zomaya. 2013. "On the Characterization of the Structural Robustness of Data Center Networks." *IEEE Transactions on Cloud Computing* 1, no. 1: 1.

Blockley, D., J. Agarwal, and P. Godfrey. 2012. "Infrastructure Resilience for High-Impact Low-Chance Risks." *Proceedings of the Institution of Civil Engineers: Civil Engineering* 165, no. 6: 13–19.

Boin, A., and A. McConnell. 2007. "Preparing for Critical Infrastructure Breakdowns: The Limits of Crisis Management and the Need for Resilience." *Journal of Contingencies & Crisis Management* 15, no. 1: 50–59.

Brown, C., Y. Ghile, M. Laverty, and K. Li. 2012. "Decision Scaling: Linking Bottom-Up Vulnerability Analysis with Climate Projections in the Water Sector." *Water Resources Research* 48, no. 9.

Burn, D. H., H. D. Venema, and S. P. Simonovic. 1991. "Risk-Based Performance Criteria for Real-Time Reservoir Operation." *Canadian Journal of Civil Engineering* 18, no. 1: 36–42.

Canon, L.-C. and E. Jeannot. 2007. "A Comparison of Robustness Metrics for Scheduling DAGs on Heterogeneous Systems." In *2007 IEEE International Conference on Cluster Computing*, 558–567. IEEE.

Castelletti, A., S. Galelli, M. Restelli, and R. Soncini-Sessa. 2010. "Tree-Based Reinforcement Learning for Optimal Water Reservoir Operation." *Water Resources Research* 46, no. 9.

Chang, S. E., T. McDaniels, J. Fox, R. Dhariwal, and H. Longstaff. 2014. "Toward Disaster-Resilient Cities: Characterizing Resilience of Infrastructure Systems with Expert Judgments." *Risk Analysis* 34, no. 3: 416–434.

Chester, M. V. and B. Allenby. 2018. "Toward Adaptive Infrastructure: Flexibility and Agility in a Non-Stationarity Age." *Sustainable and Resilient Infrastructure* 1–19.

Culley, S., S. Noble, A. Yates, M. Timbs, S. Westra, H. R. R. Maier, and A. Castelletti. 2016. "A Bottom Up Approach to Identifying the Maximum Operational Adaptive Capacity of Water Resource Systems to a Changing Climate." *Water Resources Research* 52, no. 9: 6751–6768.

Cuny, F. C. 1991. "Living with Floods: Alternatives for Riverine Flood Mitigation." Land Use Policy, 8, 331–342.

Dawson, R. J., D. Thompson, D. Johns, R. Wood, G. Darch, L. Chapman, P. N. Hughes, G. V. R. Watson, K. Paulson, S. Bell, S. N. Gosling, W. Powrie, and J. W. Hall. 2018. "A Systems Framework for National Assessment of Climate Risks to Infrastructure." *Philosophical Transactions of the Royal Society A: Mathematical, Physical and Engineering Sciences* 376, no. 2121: 20170298.

De Bruijn, K. M. 2008. "Bepalen van Schade Ten Gevolge van Overstromingen. Voor Verschillende Scenario's en Bij Verschillende Beleidsopties (Determining Flood Damage for Different Scenarios and Policy Options)." Deltares Report, Q4345.

Denaro, S., D. Anghileri, M. Giuliani, and A. Castelletti. 2017. "Informing the Operations of Water Reservoirs over Multiple Temporal Scales by Direct Use of Hydro-Meteorological Data." *Advances in Water Resources* 103: 51–63.

Döll, P. and P. Romero-Lankao. 2017. "How to Embrace Uncertainty in Participatory Climate Change Risk Management—A Roadmap." *Earth's Future* 5, no. 1: 18–36.

Drouet, L., V. Bosetti, and M. Tavoni. 2015. "Selection of Climate Policies under the Uncertainties in the Fifth Assessment Report of the IPCC." *Nature Climate Change* 5, no. 10: 937–940.

Gilroy, K. L. and R. H. McCuen. 2012. "A Nonstationary Flood Frequency Analysis Method to Adjust for Future Climate Change and Urbanization." *Journal of Hydrology* 414–415: 40–48.

Giuliani, M. and A. Castelletti. 2016. "Is Robustness Really Robust? How Different Definitions of Robustness Impact Decision-Making Under Climate Change." *Climatic Change* 1–16.

Giuliani, M., A. Castelletti, F. Pianosi, E. Mason, and P. M. Reed. 2016. "Curses, Tradeoffs, and Scalable Management: Advancing Evolutionary Multiobjective Direct Policy Search to Improve Water Reservoir Operations." *Journal of Water Resources Planning and Management* 142, no. 2: 4015050.

Giuliani, M., A. Castelletti, R. Fedorov, and P. Fraternali. 2016. "Using Crowdsourced Web Content for Informing Water Systems Operations in Snow-Dominated Catchments." *Hydrology and Earth System Sciences* 20, no. 12: 5049–5062.

Giuliani, M., D. Anghileri, A. Castelletti, P. N. Vu, and R. Soncini-Sessa. 2016. "Large Storage Operations Under Climate Change: Expanding Uncertainties and Evolving Tradeoffs." *Environmental Research Letters* 11, no. 3: 35009.

Giuliani, M., Y. Li, A. Castelletti, and C. Gandolfi. 2016. "A Coupled Human-Natural Systems Analysis of Irrigated Agriculture Under Changing Climate." *Water Resources Research* 52, no. 9: 6928–6947.

Grafton, R. Q., J. Horne, and S. A. Wheeler. 2016. "On the Marketisation of Water: Evidence from the Murray-Darling Basin, Australia." *Water Resources Management* 30, no. 3: 913–926.

Grafton, R. Q., M. McLindin, K. Hussey, P. Wyrwoll, D. Wichelns, C. Ringler, and S. Orr. 2016. "Responding to Global Challenges in Food, Energy, Environment and Water: Risks and Options Assessment for Decision Making." *Asia Pacific Policy Study* 3, no. 2: 275–299.

Gregersen, I. B., H. Madsen, D. Rosbjerg, and K. Arnbjerg-Nielsen. 2017. "A Regional and Nonstationary Model for Partial Duration Series of Extreme Rainfall." *Water Resources Research* 53: 2659–2678.

Guariso, G., S. Rinaldi, and R. Soncini-Sessa. 1985. "Decision Support Systems for Water Management: The Lake Como Case Study." *European Journal of Operational Research* 21, no. 3: 295–306.

———. 1986. "The Management of Lake Como: A Multiobjective Analysis." *Water Resources Research* 22, no. 2: 109–120.

Guillaume, J. H. A., M. Arshad, A. J. Jakeman, M. Jalava, and M. Kummu. 2016. "Robust Discrimination Between Uncertain Management Alternatives by Iterative Reflection on Crossover Point Scenarios:

Principles, Design and Implementations." *Environmental Modelling & Software* 83: 326–343.

Guo, J., G. Huang, X. Wang, Y. Li, and Q. Lin. 2017. "Investigating Future Precipitation Changes over China Through a High-Resolution Regional Climate Model Ensemble." *Earth's Future* 5, no. 3: 285–303.

Haasnoot, M., H. Middelkoop, A. Offermans, E. Van Beek, and W. P. A. Van Deursen. 2012. "Exploring Pathways for Sustainable Water Management in River Deltas in a Changing Environment." *Climatic Change* 115, no. 3–4: 795–819.

Haasnoot, M., H. Middelkoop, E. Van Beek, and W. P. A. Van Deursen. 2009. "A Method to Develop Sustainable Water Management Strategies for an Uncertain Future." *Sustainable Development* 19, no. 6: 369–381.

Haasnoot, M., W. P. A. Van Deursen, J. H. A. Guillaume, J. H. Kwakkel, E. Van Beek, and H. Middelkoop. 2014. "Fit for Purpose? Building and Evaluating a Fast, Integrated Model for Exploring Water Policy Pathways." *Environmental Modelling & Software* 60: 99–120.

Hall, J. W., R. J. Lempert, K. Keller, A. Hackbarth, C. Mijere, and D. J. McInerney. 2012. "Robust Climate Policies Under Uncertainty: A Comparison of Robust Decision Making and Info-Gap Methods." *Risk Analysis* 32, no. 10: 1657–1672.

Hamarat, C., J. H. Kwakkel, E. Pruyt, and E. T. Loonen. 2014. "An Exploratory Approach for Adaptive Policymaking by Using Multi-Objective Robust Optimization." *Simulation Modelling Practice and Theory* 46: 25–39.

Hashimoto, T., D. P. Loucks, and J. R. Stedinger. 1982. "Robustness of Water Resources Systems." *Water Resources Research* 18, no. 1: 21–26.

Hashimoto, T., J. R. Stedinger, and D. P. Loucks. 1982. "Reliability, Resiliency, and Vulnerability Criteria for Water Resource System Performance Evaluation." *Water Resources Research* 18, no. 1: 14–20.

Herman, J. D., P. M. Reed, H. B. Zeff, and G. W. Characklis. 2015. "How Should Robustness Be Defined for Water Systems Planning Under Change?" *Journal of Water Resources Planning and Management* 141, no. 10: 4015012.

Hoang, L. and R. A. Fenner. 2015. "System Interactions of Stormwater Management Using Sustainable Urban Drainage Systems and Green Infrastructure." *Urban Water Journal*, 13(7), 739–758. https://doi.org/10.1080/1573062X.2015.1036083.

Howard, R. A. 1966. "Information Value Theory." *IEEE Transactions on Systems Science and Cybernetics* 2, no. 1: 22–26.

Howard, R. A. and J. E. Matheson. 2005. "Influence Diagrams." *Decision Analysis* 2, no. 3: 127–143.

Hulme, M. 2016. "1.5°C and Climate Research after the Paris Agreement." *Nature Climate Change* 6, no. 3: 222–224.

Hurwicz, L. 1953. "Optimality Criterion for Decision Making Under Ignorance." In *Uncertain. Expect. Econ. Essays Honour GLS Shackle.*

Iglesias, A. and L. Garrote. 2015. "Adaptation Strategies for Agricultural Water Management under Climate Change in Europe." *Agricultural Water Management* 155: 113–124.

IPCC. 2014. "Summary for Policymakers." In *Climate Change 2014: Impacts, Adaptation, and Vulnerability. Part A: Global and Sectoral Aspects. Contribution of Working Group II to the Fifth Assessment Report of the Intergovernmental Panel on Climate Change*, 1–32. Cambridge, UK and New York, NY: Cambridge University Press.

Johnson, K. A. and J. C. Smithers. 2020. "Updating the Estimation of 1-Day Probable Maximum Precipitation in South Africa." *Journal of Hydrology: Regional Studies* 32: 1–15.

Kasprzyk, J. R., S. Nataraj, P. M. Reed, and R. J. Lempert. 2013. "Many Objective Robust Decision Making for Complex Environmental Systems Undergoing Change." *Environmental Modelling & Software* 42: 55–71.

Katz, R. W. 2010. "Statistics of Extremes in Climate Change." *Climatic Change* 100, no. 1: 71–76.

Kennedy, J. R. and N. V. Paretti. 2014. "Evaluation of the Magnitude and Frequency of Floods in Urban Watersheds in Phoenix and Tuscon, Arizona." *U.S. Geological Survey Scientific Investigations Report* 2014–5121.

Kim, Y., D. A. Eisenberg, E. N. Bondank, M. V. Chester, G. Mascaro, and B. S. Underwood. 2017. "Fail-Safe and Safe-to-Fail Adaptation: Decision-Making for Urban Flooding Under Climate Change." *Climatic Change* 145, no. 3–4: 397–412.

Kim, Y., et al. 2022. "Leveraging SETS Resilience Capabilities for Safe-To-Fail Infrastructure Under Climate Change." Current Opinion in Environmental Sustainability, 54, 101153.

Korteling, B., S. Dessai, and Z. Kapelan. 2012. "Using Information-Gap Decision Theory for Water Resources Planning Under Severe Uncertainty." *Water Resources Management* 27, no. 4: 1149–1172.

Kwakkel, J. H., M. Haasnoot, and W. E. Walker. 2015. "Developing Dynamic Adaptive Policy Pathways: A Computer-Assisted Approach for Developing Adaptive Strategies for a Deeply Uncertain World." *Climatic Change* 132, no. 3: 373–386.

———. 2016. "Comparing Robust Decision-Making and Dynamic Adaptive Policy Pathways for Model-Based Decision Support Under Deep Uncertainty." *Environmental Modelling & Software* 86: 168–183.

Kwakkel, J. H., S. Eker, and E. Pruyt. 2016. "How Robust Is a Robust Policy? Comparing Alternative Robustness Metrics for Robust Decision-Making." In *Robustness Analysis in Decision Aiding, Optimization, and Analytics*, 221–237.

Kwakkel, J. H., W. E. Walker, and V. A. W. J. Marchau. 2010. "Classifying and Communicating Uncertainties in Model-Based Policy Analysis." *International Journal of Technology, Policy and Management* 10, no. 4: 299–315.

Laplace, P. S. and P. Simon. 1951. *A Philosophical Essay on Probabilities.* Translated from the 6th French edition by Frederick Wilson Truscott and Frederick Lincoln Emory. Singapore: Springer.

Lempert, R. J. 2003. *Shaping the Next One Hundred Years: New Methods for Quantitative, Long-Term Policy Analysis.* Santa Monica, CA: Rand Corporation.

Lempert, R. J. and M. T. Collins. 2007. "Managing the Risk of Uncertain Threshold Responses: Comparison of Robust, Optimum, and Precautionary Approaches." *Risk Analysis* 27, no. 4: 1009–1026.

Lempert, R. J., D. G. Groves, S. W. Popper, and S. C. Bankes. 2006. "A General, Analytic Method for Generating Robust Strategies and Narrative Scenarios." *Management Science* 52, no. 4: 514–528.

Liao, K.-H. 2012. "A Theory on Urban Resilience to Floods—A Basis for Alternative Planning Practices." *Ecol Soc* 17: 15. https://doi.org/10.5751/ES-05231-170448.

Maier, H. R., B. J. Lence, B. A. Tolson, and R. O. Foschi. 2001. "First Order Reliability Method for Estimating Reliability, Vulnerability, and Resilience." *Water Resources Research* 37, no. 3: 779–790.

Maier, H. R., J. H. A. Guillaume, H. van Delden, G. A. Riddell, M. Haasnoot, and J. H. Kwakkel. 2016. "An Uncertain Future, Deep Uncertainty, Scenarios, Robustness and Adaptation: How Do They Fit Together?" *Environmental Modelling & Software* 81: 154–164.

Markolf, S. A., et al. 2021. "Re-Imagining Design Storm Criteria for the Challenges of the 21st Century." Cities, 109, 102981.

McDowell, G., E. Stephenson, and J. Ford. 2014. "Adaptation to Climate Change in Glaciated Mountain Regions." *Climatic Change* 126, no. 1–2: 77–91.

Meerow, S. and M. Stults. 2016. "Comparing Conceptualizations of Urban Climate Resilience in Theory and Practice." *Sustainability* 8(7), 701.

Miller, T. R., M. V. Chester, and T. A. Munoz-Erickson. 2018. "Rethinking Infrastructure in an Era of Unprecedented Weather Events." *Issues in Science and Technology* 34, no. 2.

Morgan, M. G., M. Henrion, and M. Small. 1990. *Uncertainty: A Guide to Dealing with Uncertainty in Quantitative Risk and Policy Analysis*. Cambridge, UK: Cambridge University Press.

Park, J., et al. 2013. "Integrating Risk and Resilience Approaches to Catastrophe Management in Engineering Systems." *Risk Anal* 33:356–367. https://doi.org/10.1111/j.1539-6924.2012.01885.x.

Paton, F. L., G. C. Dandy, and H. R. Maier. 2014. "Integrated Framework for Assessing Urban Water Supply Security of Systems with Non-Traditional Sources Under Climate Change." *Environmental Modelling & Software* 60: 302–319.

Paton, F. L., H. R. Maier, and G. C. Dandy. 2013. "Relative Magnitudes of Sources of Uncertainty in Assessing Climate Change Impacts on Water Supply Security for the Southern Adelaide Water Supply System." *Water Resources Research* 49, no. 3: 1643–1667.

———. "Including Adaptation and Mitigation Responses to Climate Change in a Multiobjective Evolutionary Algorithm Framework for Urban Water Supply Systems Incorporating GHG Emissions." *Water Resources Research* 50, no. 8: 6285–6304.

Poff, N. L., C. M. Brown, T. E. Grantham, J. H. Matthews, M. A. Palmer, C. M. Spence, and K. C. Dominique. 2015. "Sustainable Water Management Under Future Uncertainty with Eco-Engineering Decision Scaling." *Nature Climate Change* 6(1), 25–34.

Popper, S. W., C. Berrebi, J. Griffin, T. Light, and E. Y. Min. 2009. *Natural Gas and Israel's Energy Future: Near-Term Decisions from a Strategic Perspective*. Rand Corporation.

Ravalico, J. K., G. C. Dandy, and H. R. Maier. 2010. "Management Option Rank Equivalence (MORE)—A New Method of Sensitivity Analysis for Decision-Making." *Environmental Modelling & Software* 25, no. 2: 171–181.

Ravalico, J. K., H. R. Maier, and G. C. Dandy. 2009. "Sensitivity Analysis for Decision-Making Using the MORE Method—A Pareto Approach." *Reliability Engineering & System Safety* 94(7), 1229–1237.

Ray, P. A., D. W. Watkins Jr., R. M. Vogel, and P. H. Kirshen. 2013. "Performance-Based Evaluation of an Improved Robust Optimization Formulation." *Journal of Water Resources Planning and Management* 140, no. 6: 4014006.

Reliability Engineering and System Safety. 94, no. 7: 1229–1237.

Roach, T., Z. Kapelan, R. Ledbetter, and M. Ledbetter. 2016. "Comparison of Robust Optimization and Info-Gap Methods for Water Resource Management Under Deep Uncertainty." *Journal of Water Resources Planning and Management* 4016028.

Samsatli, N. J., L. G. Papageorgiou, and N. Shah. 1998. "Robustness Metrics for Dynamic Optimization Models Under Parameter Uncertainty." *AIChE Journal* 44, no. 9: 1993–2006.

Savage, L. J. 1951. "The Theory of Statistical Decision." *Journal of the American Statistical Association* 46, no. 253: 55–67.

Schneller, G. O. and G. P. Sphicas. 1983. "Decision Making Under Uncertainty: Starr's Domain Criterion." *Theory and Decision* 15, no. 4: 321–336.

Seager, T. P. 2008. "Beyond Eco-Efficiency: A Resilience Perspective." *Business Strategy Environ* 17: 411–419.

Simon, H. A. 1956. "Rational Choice and the Structure of the Environment." *Psychological Review* 63, no. 2: 129–138.

Steiner, F. R. 2006. "Metropolitan Resilience: The Role of Universities in Facilitating a Sustainable Metropolitan Future." In A. C. Nelson, B. L. Allen, and D. L. Trauger (Eds.), *Toward a Resilient Metropolis* (pp. 1–18). Alexandria, VA: Metropolitan Institute Press.

Takriti, S. and S. Ahmed. 2004. "On Robust Optimization of Two-Stage Systems." *Mathematical Programming* 99, no. 1: 109–126.

Voudouris, V., K. Matsumoto, J. Sedgwick, R. Rigby, D. Stasinopoulos, and M. Jefferson. 2014. "Exploring the Production of Natural Gas Through the Lenses of the ACEGES Model." *Energy Policy* 64: 124–133.

Wald, A. 1950. *Statistical Decision Functions*. London/New York: Chapman & Hall.

Walker, W. E., R. J. Lempert, and J. H. Kwakkel. 2013. "Deep Uncertainty." In *Encyclopedia of Operations Research and Management Science*, 395–402. Springer.

Wittholz, M. K., B. K. O'Neill, C. B. Colby, and D. Lewis. 2008. "Estimating the Cost of Desalination Plants Using a Cost Database." *Desalination* 229, no. 1–3: 10–20.

Zongxue, X., K. Jinno, A. Kawamura, S. Takesaki, and K. Ito. 1998. "Performance Risk Analysis for Fukuoka Water Supply System." *Water Resources Management* 12, no. 1: 13–30.

6

EVALUATING CRITICAL DRAINAGE INFRASTRUCTURE'S RESILIENCE TO EXTREME WEATHER

6.1 INTRODUCTION

In recent years, city municipalities and governments at various levels have increasingly recognized the importance of building resilient infrastructure. This recognition is driven by factors such as extreme weather events, often referred to as the "New Normal," attributable to climate change, primarily caused by human activities like urbanization, deforestation, and the expansion of paved areas. In the context of urban drainage infrastructure, resilience refers to the system's ability to withstand and recover from adverse situations. It involves the capacity to bounce back from challenging circumstances, pressure, or distress. The degradation of critical infrastructure, such as urban drainage systems, can be significantly exacerbated by emerging threats. These dynamic factors can collectively exert a detrimental impact on these systems, posing a significant challenge to their long-term sustainability.

The inherent uncertainty surrounding future conditions can lead to doubts about the effectiveness of both traditional and innovative methods in responding to extreme weather events. This skepticism stems from the uncertain magnitude and scope of the long-term consequences of those incidents. It is crucial to incorporate at least minimal adaptive enhancements into urban drainage infrastructure to mitigate vulnerability to catastrophic flooding, casualties, ecological damage, and economic losses, regardless of whether these events are classified as ordinary or extraordinary. An urban drainage system is expected to demonstrate reliability by minimizing instances of failure and ensuring a satisfactory level of service for most of its operational duration. Furthermore, an urban drainage infrastructure system is anticipated to exhibit resilience by effectively reducing the duration and severity of expected failures, including combined sewer overflows (CSOs), water quality issues, and flooding events.

6.2 RISK-OF-FAILURE CATEGORIZATION

Climate change is expected to increase precipitation levels in many regions. Traditional risk analyses involve identifying, characterizing, and categorizing drainage infrastructure that could potentially fail due to extreme weather events. In this context, the term *risk of failure* refers to any event or circumstance that could compromise the system's ability to provide a predetermined level of service. Various terms, such as risk, event, perturbation, disturbance, shock, and crisis, are used in the literature. These terms are typically used in traditional risk management methodologies, where the focus is on assessing the likelihood of their occurrence and the resulting consequences. The classification of failure risk may vary depending on the specific context or system under consideration. This work provides a succinct overview of four distinct categories of failure risk: technical, environmental, sociocultural, and health:

1. *Technical*: Potential outcomes may include various adverse effects on drainage infrastructure components. These can range from malfunctions—in pipes, inlets, and catch basins due to excessive stormwater runoff discharge or volume—to overcapacity. Damage to or reduced efficiency in pumping operations may also occur. Other potential issues include landslides, street failure, erosion, financial losses, and the need for repairs or replacements due to property losses. Awareness of these potential

consequences and the implementation of appropriate preventive and mitigative measures are crucial for the smooth operation and preservation of infrastructure and resources. In urban environments, numerous variables can significantly influence the capacity of the drainage network and other elements of the drainage infrastructure system.

2. *Environmental*: The dispersion of a broad array of contaminants and hazardous substances, such as oil, grease, heavy metals, pharmaceuticals, pesticides, and fertilizers, from various sources, including streets, parking lots, and agricultural land, into water, soil, and air can profoundly impact ecosystems and species. Economic considerations should include the financial implications of damages, the costs associated with environmental remediation, and any additional expenses that may arise. Secondary costs may include potential disruptions to job performance resulting from infrastructure failures, such as those affecting roads, railways, and internet connectivity. Environmental threats, typically characterized by their gradual onset, may include the potential proximity of infrastructure to water bodies and the occurrence of urban creep. These events are often expected and/or predictable.

3. *Socio-cultural*: Socio-cultural factors can lead to variations in the extent of damage and pollution across different regions within a city, municipality, or country. In areas populated by economically disadvantaged individuals, environmental damage and pollution can contribute to the development of social stratification or class distinctions. Social risk threats refer to any external force, entity, or actor outside the system, which can include factors such as the natural or built environment, proximity to other infrastructure, and distance to critical infrastructures like streets, buildings, and traffic.

4. *Health*: The harmful effects of environmental degradation and pollution may cause illness, injury, or even death. Moreover, the quality of drinking water can significantly impact public health. An acute health and safety risk refers to a situation posing

immediate and significant dangers, including potential impacts on biodiversity, loss of life and property, and intangible consequences such as increased anxiety.

This categorization aims to enhance the identification of potential failure risks in different scenarios, thus providing valuable insights to decision makers such as municipal authorities, policymakers, and regulators. Partitioning the *risk-of-failure space* can facilitate a more targeted analysis methodology and aid in the prioritization of interventions. This is crucial as a new paradigm is necessary to effectively address extreme weather challenges.

6.3 RISK-OF-FAILURE SPACE DEVELOPMENT

The methodology for risk-of-failure space development employs a comprehensive and strategic analysis to identify and assess risks in complex urban drainage infrastructure systems. The primary objective is to tackle the challenges presented by adverse weather conditions efficiently and cost-effectively while enhancing the safety and efficiency of these systems. The development of the risk-of-failure space is crucial in understanding the fundamental nature of risk and its evaluation. The risk-of-failure space should consider the complexity of the challenges, including:

1. The scale of the infrastructure challenges
2. The orientation of major stakeholders and agencies toward traditional responses
3. The duration of the challenges relative to the solution
4. The interconnectedness of essential services
5. The cumulative and direct impacts, as well as indirect impacts
6. The politics of a highly securitized operating environment
7. The significant shortcomings given gaps in evidence and analysis
8. The challenges associated with enforcing and applying design standards
9. Funding that does not align with the duration or scale of the needs

6.4 BASIC CRITICAL DRAINAGE AREA RESILIENCE ASSESSMENT

A resilient urban drainage system is defined by its ability to withstand severe weather events, maintain functionality under such conditions, and recover promptly from their impacts. With climate change, the primary concern for many urban areas is the intensification of the hydrological cycle, characterized by increased precipitation and extreme weather events. These events can lead to flooding, CSO, and reduced flow capacity within the system due to increased levels of piping infiltration. Evaluating resilience in critical urban drainage infrastructure is vital for professionals in various fields, such as policymakers, engineers, and planners. The main goal of this assessment is to identify and delineate specific zones within drainage areas that are susceptible to the impacts of extreme weather. The first step in assessing overloaded subareas involves simulating the existing drainage infrastructure using a standardized methodology commonly used in urban drainage mapping projects within developing regions:

1. *Defining an assessment boundary*: This process involves active engagement with local stakeholders to establish a specific area of interest for evaluation.
2. *Establishing an assessment protocol*: Developing a comprehensive drainage assessment protocol involves defining and categorizing various drainage characteristics and analysis parameters, along with conducting boundary and topological surveys.
3. *Collecting data*: The objective is to systematically document the location and condition of the drainage infrastructure within a defined area of interest. This includes gathering information on historical flood records, as well as the diameter, length, invert, material, laying year, soil conditions, coordinates, and joint type of each drainage pipe in the area.
4. *Analyzing the overall cost*: This includes performing a cost-benefit analysis (CBA), which forms the basis for prioritizing potential adaptation strategies. One of its advantages is that it enables the comparison of various impacts using a unified metric,

which is crucial for accurately allocating costs and benefits and assessing their overall values. However, incorporating accurate estimations of intangible assets, which hold value but are not easily quantifiable in market terms, can present a significant challenge in CBA.

5. *Initiating drainage mapping*: This involves commencing drainage mapping and implementing a comprehensive assessment of the resilience of critical drainage areas. The development of drainage master plans currently utilizes similar steps. Given current extreme weather conditions, additional simulations are necessary to capture the dynamic changes in the discharge of drainage systems. This will facilitate the documentation of potential future scenarios that may emerge and require careful consideration (see Figure 6.1).

Upon analyzing the simulation results, which include both the present state and potential future scenarios, practitioners can assess the need for adaptation. Three parameters—pipe capacity assessment, extent of flooding, and flooding volume—can be used to determine whether modifications or upgrades to an existing drainage system or structure are needed to enhance its functionality, efficiency, or compliance with new drainage design standards:

1. Weighted pipe capacity assessment involves evaluating the maximum discharge of each drainage main and comparing it to the maximum design discharge. This assessment accounts for nonstationary hydrologic processes to accommodate extreme weather events and hydrologic extremes induced by climate change. This parameter allows for the evaluation of the pipe capacity of the entire catchment area, also known as a watershed or drainage basin, or specific sections thereof. It can be calculated using Equation 6.1:

$$\text{Weighted pipe capacity} = \frac{\sum_1^n \frac{Q_{nonstationary,x}}{Q_{design,x}} \times l_x}{\sum l}, \tag{6.1}$$

1. Defining an Assessment Boundary:
- Engage with local stakeholders to establish a specific area of interest for evaluation
- Organize an interdisciplinary professional team

2. Establishing an Assessment Protocol:
- Develop a comprehensive drainage assessment protocol
- Define and categorize various drainage characteristics, analysis parameters, scenarios, risk, vulnerability, and conduct boundary and topological surveys

3. Collecting Data:
- Systematically document the location and condition of the drainage infrastructure within a defined area of interest and include information on historical flood records

4. Analyzing Overall Cost:
- Perform a cost-benefit analysis (CBA)
- Evaluate alternatives

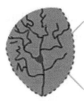

5. Initiating Drainage Mapping:
- Start the process of drainage mapping and conduct a thorough evaluation of the resilience of crucial drainage areas

Figure 6.1 Commonly used steps in urban drainage mapping projects within developing regions

where $Q_{nonstationary}$ is the calculated maximum discharge (m³/s) using nonstationary processes, Q_{design} is the standard calculated maximum discharge (m³/s), and Σl is the summation of the length of the pipe (m).

2. The manhole inundation assessment ratio is employed to analyze the degree of flooding in manholes. This metric, representing the ratio of flooded manholes to the total number of manholes in a given area, provides insight into the extent of flooding and its overall impact on manhole infrastructure. This parameter can be assessed for the entire system or its individual components and can be computed using Equation 6.2:

$$\text{Manhole inundation assessment} = \frac{\Sigma MH_{Inundated}}{\Sigma MH_T},\qquad(6.2)$$

where $MH_{inundated}$ is the total number of manholes inundated or flooded and ΣMH_T is the total number of manholes for the evaluated drainage area.

3. Specific flooding volume refers to the amount of stormwater runoff that can cause flooding in a given catchment area. A flood volume of 10 cubic meters per hectare can be interpreted as the volume of water that accumulates during a flood event, equivalent to an effective rainfall of 1 millimeter or 1 liter per square meter. This flood volume can be calculated using Equation 6.3:

$$\text{Specific flood volume } = \frac{\Sigma F_v}{\Sigma A},\qquad(6.3)$$

where F_v represents the flood volume, which is directly related to the calculated flood volume and the catchment area, denoted as A.

The inclusion of these three primary parameters allows for an initial categorization of the necessary adjustments to the drainage infrastructure due to extreme weather events. This categorization is based on predefined threshold values assigned to each assessed drainage basin.

Table 6.1 presents these values used to initially categorize the requirements of drainage infrastructure for adaptation, retrofitting, and construction of new resilient infrastructure systems. Implementing these measures is crucial for cities and communities to effectively mitigate the increasingly frequent natural hazards resulting from climate change and to prepare for future occurrences.

Table 6.1   Predetermined threshold values

	Weighted Pipe Capacity	Manhole Inundation	Specific Volume
	0	0	0
	0.2	0.05	2
None/Low	0.4	0.1	4
	0.6	0.15	6
	0.8	0.2	8
Medium	1	0.25	10
	1.2	0.3	12
	1.4	0.35	14
High	1.6	0.4	16
	1.8	0.45	18
	2	0.5	20

6.5 INDICATORS

Indicators serve various functions, including enhancing decision-making processes and facilitating efficient actions, by providing policymakers with streamlined, consolidated information. They can be classified based on system performance, capacity exceedance, and potential consequences of capacity exceedance. These classifications can also be categorized based on their relation to system events, including pre-, intra-, and post-event. Indicators integrate knowledge from physical and social sciences in the decision-making process, enable assessment and adjustment toward achieving sustainable development goals, offer timely alerts to mitigate potential economic, social, and environmental setbacks, and effectively convey ideas, thoughts, and values. Three indicators are particularly relevant to climate change:

1. Reliability indicators
   - Hydraulic reliability index
   - Temporal reliability index
   - Volumetric reliability index
2. Resilience indicators
3. Sustainability indicators

6.5.1 Reliability Indicators

Reliability, technically, measures a system's ability to minimize service failures within its expected operational lifespan under typical operating conditions—typically measured against standard loading criteria. Service failure refers to the inability to meet regulatory standards, emphasizing long-term planning for urban drainage systems in uncertain future conditions.

Reliability indicators are determined by the consistency in delivering satisfactory service, measured by the probability of no system failures within a specified period (e.g., a year) that does not exceed a predetermined threshold. The mathematical formula for reliability indicators can be expressed via Equation 6.4:

$$Reliability = 1 - \sum_i \frac{\mathcal{F}_i}{T}, \tag{6.4}$$

where $\mathcal{F}_i$ represents the combined measure of the duration and intensity of each individual failure during the evaluated period T, which is on an annual basis. Reliability refers to the duration of uninterrupted system operation without any failures within a given year.

6.5.1.1 Hydraulic Reliability Index

The hydraulic reliability index is crucial in assessing the reliability of the hydraulic systems of a drainage network (see Equations 6.5 to 6.7):

$$\text{Hydraulic reliability index} = \frac{1}{N}\sum_{t=1}^{N} Z,\qquad (6.5)$$

$$Z_t = 1 \forall\, X_t \in S,\qquad (6.6)$$

$$Z_t = 0 \forall\, X_t \in F.\qquad (6.7)$$

In the evaluation of a drainage system over a specific time interval, the existing condition (or current state) and the associated value are denoted by Z_t and X_t, respectively. Satisfaction, denoted as S, signifies that the drainage system is functioning efficiently and meeting the specified criteria. A *satisfactory state* is a condition where the drainage network effectively manages stormwater runoff without being overwhelmed, ensuring that the discharge or volume of the runoff does not exceed the system's capacity. Conversely, a failure state, denoted as F, indicates that the drainage network is not functioning as intended and fails to meet the necessary criteria. Failure states occur when the depth of the stormwater runoff exceeds the capacity of the drainage system components throughout the system. N represents the total number of time intervals under analysis, enhancing the evaluation of the drainage system's efficiency within a specified period.

6.5.1.2 Temporal Reliability Index

In the context of drainage infrastructure systems, the temporal reliability index metric serves as a tool to evaluate the reliability and consistency of a drainage system's operational performance within a specific catchment area over a given time period. This analysis yields insights into the performance of the drainage network based on specific measurements or observations recorded over time. The temporal reliability index can be computed using Equation 6.8:

$$\text{Temporal reliability index} = 1 - \left(\frac{1}{T}\sum_{f=1}^{n} d_f\right),\qquad (6.8)$$

where n, f, d_f, and T are used in the calculation of the cumulative count of drainage systems that showed inadequate performance (failures) or experienced surcharging. n denotes the total number of systems that have shown inadequate performance or surcharging, while f represents the count of instances of inadequate performance (failures) or surcharging, thereby tracking the number of system failures. d_f denotes the duration during which the system remains in an inadequate performance or surcharging mode f. This variable measures the length of time the system stays in each failure mode. T represents the duration of the operating period, helping to determine the overall time frame during which the system is operational. By understanding and utilizing these variables, decision makers can effectively analyze and evaluate the operational performance of the drainage system within a specific catchment area over time.

In addition to the previously discussed metrics, it is important to note that there are various other reliability indices available for assessing the operational performance of a drainage system in a particular catchment area over time. These include volumetric reliability indices, which evaluate system reliability based on their capacity to manage designed stormwater volumes. Another option is the reliability index of the CSOs, which assesses the reliability of the combined sewer system in managing and preventing overflow events. Drainage component reliability indices evaluate the reliability of individual components within drainage systems. These alternative indices provide valuable insights into the performance and reliability of different drainage system components. When these metrics are taken into consideration by decision makers, they provide valuable insights that can inform the formulation of operational strategies prioritizing efficiency, security, and reliability in the drainage system.

6.5.2 Resilience Indicators

To effectively address future challenges, decision makers must prioritize both fail-safe reliability and safe-to-fail resilience in drainage solutions. These systems have the capacity to demonstrate enhanced flexibility and expedite recovery, minimizing damage and mitigating service disruption in the event of a failure. Resilience is defined as the system's ability to mitigate the impact of exceptional conditions, such as threats or a combination of threats, by reducing the magnitude and duration of service failures. This definition encompasses all previously mentioned characteristics.

The assessment of failure events involves considering both the magnitude and duration of such events within the evaluated time frame. Resilience indicators can generally be expressed using Equation 6.9:

$$Severity = \sum_i \frac{d_i \; X \; m_i}{T},$$ (6.9)

where m_i and d_i denote, respectively, the magnitude and duration of failures within the assessed period T, which corresponds to a year. Equation 6.9 is not designed to provide an exact measurement of the absolute severity of failures. Instead, it allows for a comparison of the relative failure severity of various options within the context of consistent annual conditions.

6.5.3 Sustainability Indicators

The operational performance of urban drainage systems, in the context of future changes, is characterized by the adopted definitions of reliability and resilience. These attributes describe the system's resilience, including its ability to withstand, adapt to, and recover from stress and failure, thereby mitigating potential consequences. Sustainability, defined as "the extent to which the system maintains service levels over the long term while simultaneously maximizing social, economic, and environmental objectives," is associated with the system's long-term performance. This includes both periods of failure and nonfailure and extends beyond the system's designated lifespan. The consequences of failure, particularly in terms of the system's reliability and resilience, that impact broader social, environmental, and economic systems in urban areas, are factors influencing the system's sustainability.

Operational impacts, such as the magnitude or duration of flooding events, interact with the three pillars of sustainability (social, environmental, and economic), resulting in various consequences. These consequences pose threats to the recipients of water services. There-fore, sustainability indicators relate to the potential effects of failure on society, the economy, and the environment.

The broader implications include four additional objectives that are not considered in the reliability and resilience indicators—river flood-ing, greenhouse gas emissions, cost, and acceptability. The absence of reliability and resilience underscores the operational aspect of these factors since there is no apparent connection between these additional goals and the failure of operational performance. These objectives are consequential, impacting not only the operational performance of the system but also playing a crucial role in assessing the long-term economic, environmental, and societal consequences associated with investment decisions initially driven by operational factors, such as flooding or water quality.

6.6 RELIABILITY, RESILIENCE, AND SUSTAINABILITY ROBUSTNESS INDICES

Analyzing the operational performance of drainage infrastructure us-ing robustness indices and examining their correlation with reliability, resilience, and sustainability is crucial. Robustness indices are instru-mental in assessing the effectiveness of drainage systems under various conditions. Reliability, the consistent performance of a system with-out failure, is a key aspect of any system. Robustness indices offer a quantitative measure of this reliability, enabling an evaluation of the system's ability to handle disruptions without performance compro-mise. Resilience pertains to the system's capacity to maintain function-ality amidst disruptions, as assessed by the aforementioned indices. Sustainability, a comprehensive term, aims to balance environmental, social, and economic factors in resource utilization.

This necessitates data collection on the system's response to varying disturbance levels. Strategy robustness assessment involves evaluating regret, the relative performance loss across all objectives, and future scenarios previously analyzed. Ignoring uncertainty can lead to inef-fective strategies that fail to address potential threats or capitalize on opportunities associated with increased uncertainty. Performance decline will be assessed by analyzing each performance indicator, providing insights into the negative impacts resulting from performance failures. The concept of regret, or opportunity loss, originally introduced by Leonard Savage in 1951, can effectively provide decision recommendations for mutually exclusive strategies. The regret of a strategy is the difference between its performance for a specific objective and the performance of the best strategy for the same future scenario and objective.

Savage's concept of regret, also known as opportunity loss, has been pivotal in suggesting mutually exclusive alternative strategies. This criterion aims to minimize the loss of potential opportunities and can be measured by comparing the performance of a strategy against the optimal strategy in a given scenario and objective. Performance disparity, a metric used to assess strategy effectiveness, identifies missed opportunities or suboptimal results. Using the concept of regret, decision makers evaluate the impact of a strategy on the anticipated future condition. This concept can be mathematically represented as Equation 6.10:

$$
\begin{aligned}
Regret(p_s, f_s, j) = [&Performance(O_s, f_s, j) \\
&- Performance(p_s, f_s, j)],
\end{aligned}
$$

$$(6.10)$$

where p_s is the performance of a strategy, O_s is the optimal strategy, f_s is the given scenario, and j is the objective or target.

Savage also emphasized the potential for post-decision regret among decision makers, which arises from the realization of event outcomes, known as states of nature. Thus, it is recommended that decision makers adopt strategies aimed at minimizing regret before making a final choice among available alternatives, as indicated in Equation 6.11:

$$
\begin{aligned}
Regret_j(p_s, f_s) = \big|&max_{O_s}[Performance_{j(O_s, f_s)}] \\
&- Performance_j(p_s, f_s)\big|.
\end{aligned}
$$

$$(6.11)$$

Regret-based approaches operate on the principle of minimizing opportunity loss or regret. They select the strategy that best addresses all objectives across various potential future scenarios. This principle provides a decision-making framework aimed at maximizing favorable outcomes and minimizing potential regrets. Evaluating the reliability, resilience, and sustainability of a system in various future scenarios necessitates considering a diverse set of indicators and objectives. Some indicators are evaluated operationally, involving periodic assessments to monitor the system's continuous performance. However, certain metrics, such as costs, are assessed throughout the system's entire life cycle, accounting for any potential variations or fluctuations over time. Analyzing these indicators provides a comprehensive understanding of the system's performance and financial implications.

6.7 TRADE-OFF BETWEEN COST, RESILIENCE ENHANCEMENT, AND OPTION IDENTIFICATION

Trade-offs are integral to decision-making processes. Decision-making scenarios often require the careful consideration of multiple options. In such cases, individuals or organizations must assess the benefits and drawbacks of these options, as well as the trade-off between cost and resilience improvement. Implementing resilience measures often entails a significant financial investment. Allocating resources toward enhancing resilience requires careful consideration since it may lead to increased expenses. However, choosing a more budget-friendly alternative may compromise durability. It is crucial to evaluate the trade-off between cost and resilience enhancement. Cost refers to the monetary expenses associated with a particular choice or action, while resilience enhancement refers to measures taken to increase the capacity to withstand and recover from various challenges or disruptions. To make a more informed assessment of the advantages and disadvantages of additional tasks, such as training, monitoring, collaboration, and maintenance, it is necessary to understand the trade-off between the cost and resilience improvement of stormwater drainage infrastructure.

In this section, we explore four commonly used techniques for evaluating the balance between cost, resilience enhancement, and option identification. These techniques include maintaining the status quo or implementing the minimum necessary measures, conducting a CBA, performing a cost-effectiveness analysis (CEA), and utilizing multi-criteria analysis (MCA). These techniques are widely used across various industries to make informed decisions and achieve a balance between these crucial factors.

6.7.1 Take No Action/Do Nothing, Do the Minimum, or Do Something

The *take-no-action* option, also known as the *do-nothing* option, serves as a benchmark for evaluating the impact of alternative options concerning the current situation. It aims to determine whether the alternatives under consideration offer improvements or disadvantages compared to the existing state. The *current state* alternative represents a zero-investment scenario, where no additional costs are incurred beyond the existing expenditures on operations, maintenance, and related activities. Analyzing current infrastructure condition data is crucial to establish baseline conditions.

Evaluating the do-nothing alternative typically begins with the collection of historical data, including information on infrastructure assets, pipe capacity, vulnerability to flooding, lifespan, maintenance records, and other relevant factors. Practitioners must conduct a comprehensive assessment of infrastructures, considering various aspects such as geographical placement, lifespan, maintenance history, performance, and operational challenges. This assessment should provide a detailed depiction of both current and anticipated requirements, expenses, materials, and other relevant factors to facilitate comparative analysis.

The *do-the-minimum* option refers to a scenario where a project is executed with minimal effort and cost. This alternative assumes the occurrence of specific, negligible investment expenditures that exceed current operational and maintenance expenses. For example, partially modernizing an existing infrastructure requires less investment effort and expenditure than a complete modernization. The do-the-minimum option provides a solution that incurs the lowest cost to meet the overall objectives or requirements. This option includes the associated incremental costs and benefits, such as employment opportunities, which can be used for comparison purposes.

The *do-something* option necessitates action. Decision makers must identify potential countermeasures to do-nothing and do-the-minimum alternatives. Do-something solutions are selected because they best meet the goals or specifications, depending on the investment required by the do-something alternatives. In many situations, cost is the primary consideration, and each alternative is evaluated based on the amount of savings and level of investment. The do-something option(s) should be detailed, along with any incremental expenses and benefits (e.g., costs, jobs, or other comparison factors).

6.7.2 Cost-Benefit Analysis

In the field of drainage infrastructure management, CBA is a frequently employed tool for the effective evaluation of adaptation alternatives. This method is used to assess the feasibility of various adaptation options, such as repair, rehabilitation, retrofitting, or even the construction of new drainage systems to enhance the operational efficiency of existing infrastructure or to implement new projects. CBA involves a systematic assessment and calculation of the costs and benefits associated with each alternative, expressed in monetary units, to facilitate informed decision making and selection of the most beneficial alternatives.

The CBA framework forms the basis for prioritizing potential adaptation strategies. One of its advantages is that it enables the comparison of various impacts using a unified metric, which is crucial for accurately allocating costs and benefits and assessing their overall values. However, incorporating accurate estimations of intangible assets, which hold value but are not easily quantifiable in market terms, can present a significant challenge in CBA. This could lead to the exclusion of nonmarket costs and benefits, potentially resulting in misleading outcomes in the analysis.

6.7.2.1 Evaluating Adaptation Alternatives Using Cost-Benefit Analysis

In conventional CBA investigations, anticipated costs and benefits must be evaluated to develop more comprehensive approaches for valuing benefits, which can then be integrated into planning studies. However, certain benefits may not be easily measurable or quantifiable. Incorporating CBA alongside other nonmonetary assessments of benefits can enhance flexibility in strategic planning and rationalize resource allocation to innovative projects or activities. This approach acknowledges the challenges associated with quantifying the financial value of certain benefits. The following are procedural guidelines for employing CBA in the assessment of adaptation alternatives:

1. *Objective and options for adaptation*: It is imperative for decision makers to reach a consensus on the objective of adaptation and subsequently investigate various potential adaptation options. The adaptation objective should be defined concisely and unambiguously, and its achievement should be quantifiable in monetary terms. One possible implication of this concept is the reduction of vulnerability, especially by attaining a specified level of protection against the risks posed by extreme weather events due to climate change, such as floods.

2. *Establishment of a baseline*: The creation of a baseline is crucial for effectively evaluating the costs and benefits of implementing an adaptation intervention. The baseline represents the current state of affairs without any intervention, while the project line indicates the anticipated outcome from the successful execution of the adaptation option. By analyzing these two scenarios comprehensively, decision makers can accurately assess the financial implications and benefits associated with the implementation. For example, historical data on disaster frequencies and resulting damages can be used for the baseline vulnerability assessment. Calculations incorporating various data and model outputs can facilitate the comparison between

a scenario with a specific condition and one without it, thereby avoiding the common mistake of comparing situations before and after the implementation of the said condition.

3. *Quantification and aggregation*: This step involves calculating and consolidating expenses within specified time intervals. The costs associated with an adaptation action include both direct costs, such as investments and regulatory expenses, and indirect costs, such as social welfare losses and transitional costs. Cost evaluations for municipal operations and infrastructure planning primarily focus on the current costs of existing infrastructure and program activities, as well as future costs of upcoming projects. Estimates should effectively account for the dynamic nature of activity costs over time and the impact of inflation rates. Therefore, benefits should also be quantified and aggregated within designated time periods. The benefits of implementing an adaptation intervention should include the reduction of damages caused by climate change impacts and any relevant co-benefits.

4. *Comparative analysis of total costs and benefits*: In this section, *aggregated costs* refer to the total expenditures related to a specific project, program, or investment. Conversely, *aggregated benefits* refer to the total advantages that could result from a specific initiative, policy, or investment. A comprehensive analysis and evaluation of both costs and benefits are essential for informed decision making and an accurate assessment of the overall value associated with our objectives and priorities. By considering the total costs and potential benefits, we can gain a thorough understanding of the financial implications and potential gains of a specific project. The present value of project costs is determined by aggregating expenses over a specified time frame and applying a discount rate, as calculated using Equation 6.12:

$$P_v = \sum_{y=0}^{n} \left(\frac{C_y}{(1+r)^y} \right), \quad\quad (6.12)$$

where y and r represent the variables for the number of periods and the rate of return or discount, respectively. The values for y range from 0 to the projected final year of operation. Decision makers and planners must consider three indicators when assessing the efficacy of their decisions; these indicators are valuable tools in assessing the efficiency of their choices:

a. *Net present value*: This is a financial metric that gauges the profitability of an investment or project. Net present value (NPV) is a useful tool for analyzing potential returns on an investment and making informed decisions about its viability. The mathematical formula for calculating NPV is given by Equation 6.13:

$$N_{pv} = \sum_{y=0}^{n} \left(\frac{B_y}{(1+r)^y} - \frac{C_y}{(1+r)^y} \right), \qquad (6.13)$$

where B_y represents the cost benefits and C_y represents the costs. n option is considered acceptable if the NPV exceeds zero. NPVs can be used to determine the relative importance of different investment opportunities for effective fund allocation.

b. *Benefit-cost ratio*: This quantitative metric evaluates the relative benefits of a project or investment in relation to its associated costs. The assessment of benefits and costs is typically conducted within a specified time frame and often involves the use of discounting techniques. Discounting recalculates the value of future benefits and costs to reflect their present worth, acknowledging the time value of money. Therefore, when comparing benefits and costs occurring at different time intervals, it is crucial to consider specific variables. The benefit-cost ratio is widely used in various fields, such as economics, finance, and public policy, to facilitate informed decision making regarding resource allocation and project prioritization.

c. *Internal rate of return*: The internal rate of return (IRR) is a popular financial indicator for assessing the profitability of a project or investment. During the evaluation of various alternatives, it is essential to consider the IRR as a key element in assessing the comparative merits of each option. It is generally more advantageous to select the option with a higher IRR because a higher IRR indicates a greater potential for generating positive returns and achieving profitability. In other words, an option with a higher IRR is more financially appealing and advantageous.

6.7.2.2 Incorporating the Project Lifespan in Cost Estimations Using Discount Rates

When estimating project costs, decision makers must consider the entire lifespan of the project. This involves evaluating NPV of anticipated benefits and expenses over the project's lifetime using discount rates. Discount rates can be converted into discount factors, which include both the rate and year. This conversion is achieved using Equations 6.14 and 6.15, respectively:

$$D_f = \frac{1}{(1 + \mathcal{R})^y},$$ (6.14)

where D_f is the discount factor: $(D_f < 1)$, and

$$F_v = P_v - (1 + \mathcal{R})^y \text{ or } P_v = F_v * (1 + \mathcal{R})^{-y},$$ (6.15)

where F_v is the future value, P_v is the present value, and y is the year.

Discount and inflation rates are used to incorporate the dynamic nature of monetary value into analytical assessments. The discount rate allows us to determine the future value of money by considering its present value. Conversely, inflation rates can be used to project the potential future nominal value of a given amount of money. To establish the relationship between the present and future value of money, one can use a discount rate, represented as R in Equation 6.15.

6.7.2.3 Strengths and Weaknesses of Cost-Benefit Analysis

CBAs are highly advantageous, given their ability to aggregate and compare various types of benefits or costs into a single monetary amount. However, a limitation of CBAs is their dependence on quantifying all benefits in monetary terms, with a focus on efficiency. CBAs may not adequately consider the equity implications of distributing the costs and benefits of adaptation measures among different stakeholder groups. This includes neglecting an assessment of the affordability for those who stand to benefit from the policy under consideration.

The claim that the projects or policies with the highest benefit-cost ratio are socially desirable rests on the assumption that the beneficiaries can potentially compensate those who are adversely affected while still achieving overall improvement. However, whether such compensation occurs depends on the formulation of the adaptation policy. An additional challenge with CBAs is the requirement to assign monetary values to various costs and benefits that occur at different time intervals. This requires discounting future costs and benefits to calculate their present value, which involves the challenge of selecting an appropriate discount rate.

6.7.3 Cost-Effectiveness Analysis

CEA is a widely used approach for assessing the relative value of various interventions or policies. It involves a comprehensive evaluation of their costs and outcomes. In CEA, a single measure of effectiveness is compared with a corresponding cost. This comparison is essential in policies such as those for evaluating flood risk mitigation projects. The standard practice involves generating a CEA as in Equation 6.16:

$$CEA = \frac{E}{C},\qquad(6.16)$$

where E is the effectiveness measure and C is the cost variable. When deciding to utilize the CEA methodology, decision makers are advised to:

1. Ensure the appropriate output or outcome is used to construct the ratio accurately. Experimenting with different output variations is recommended to assess any significant impacts on

the comparative rankings of the options and programs under consideration.

2. Begin by compiling a comprehensive inventory of all potential costs and benefits, even if certain items may be subsequently excluded. Omitting relevant factors during the analysis may result in the exclusion of critical elements.

3. Conduct a sensitivity analysis to assess the robustness of the CBA outcome when key variables are altered. Information on costs, benefits, and risks is often uncertain, especially when considering future scenarios.

The CEA provides guidance for selecting the most suitable policy or project among several alternatives. In the context of policy evaluation, the CEA framework can assess and rank various policy options, which may not be mutually exclusive. Multiple policies can be implemented simultaneously, but at least some of these policies must be implemented to address the issue at hand. Among potential policies $i = 1, 2, 3 \ldots\ldots n$, each with corresponding costs C_i and effectiveness E_i, CEA can rank the policies based on their attributes (see Equation 6.17):

$$CER_i = \frac{E_i}{C_i}. \qquad (6.17)$$

This ranking can be utilized to choose projects that align with the allocated budget, as shown in Equation 6.18:

$$\text{Rank by } CER_i \qquad \Sigma_i C_i = C. \qquad (6.18)$$

6.7.4 Multi-Criteria Analysis

MCA is a decision-making technique that involves evaluating and comparing multiple criteria or factors. Like CEA, MCA does not provide conclusive evidence regarding the feasibility of implementing an investment project or policy. MCA evaluates efficiency and cost in a ratio format. Its scope is limited to the selection of options within a portfolio where certain options are required. Both MCA and CEA are considered effective in optimizing the relationship between effectiveness and cost. However, they may be perceived as lacking economic efficiency. In certain circumstances, MCA yields results comparable to those obtained via CBA.

MCA typically involves the use of carefully chosen scores and weights determined by subject matter experts. MCA exhibits a lower level of accountability than CBA, as the monetary units in MCA are derived from expert preferences rather than individual preferences. The basic equation (see Equation 6.19) for calculating the final score of an investment project or policy using the most basic form of MCA is as follows:

$$S_i = \sum_i w_i * S_{ij}, \tag{6.19}$$

where i is the ith option, j is the jth criterion for selection, w is the weight, and S is the score.

In its simplest form, MCA determines the final result by calculating a weighted average of the scores. The option with the highest weighted score is deemed the most favorable. More advanced methodologies can be employed to address complex decision-making scenarios. MCA is often recognized for its increased transparency compared to CBA, primarily due to its explicit statement of objectives and criteria, rather than relying on underlying assumptions. However, MCA may exhibit less transparency compared to CEA because MCA incorporates multiple objectives, while CEA focuses on a single objective. The clarity in addressing issues related to time discounting and adjustment of relative valuations in marginal cost analysis is often ambiguous.

Distributional implications are frequently a primary focus in MCA, allowing for the explicit integration of distributional effects within the MCA framework. MCA shares several similarities with CEA, which also considers multiple indicators of effectiveness. Technically, CEA is more similar to straightforward MCA models due to the need to normalize various effectiveness indicators measured in different units. This normalization process involves converting units into scores and then aggregating them through a weighting procedure.

6.8 CHAPTER SUMMARY

The operational performance of an urban drainage system, both before and after a failure, is used to define its resilience and reliability. The resilience of an urban drainage system refers to its ability to withstand, adapt to, and recover from stress and failure, thereby minimizing adverse effects. Conversely, a drainage system's sustainability is determined by its ability to balance financial, social, and ecological objectives while maintaining operational functionality and service levels over time, particularly during extreme weather events, often referred to as the "New Normal."

Even when a drainage system has exceeded its lifespan, as is the case with much of the infrastructure in large cities, its sustainability is determined by its long-term performance, including periods of both failure and nonfailure. To effectively address future challenges, decision makers must ensure that drainage solutions operate securely, prioritizing both fail-safe reliability and safe-to-fail resilience to the greatest extent possible. These systems have the ability to demonstrate enhanced flexibility and expedite recovery to minimize damage and mitigate service disruption in the event of a failure. This can be achieved using various operational function indicators for drainage infrastructure.

The use of these tools can improve decision-making processes and facilitate efficient actions by streamlining, clarifying, and providing policymakers with consolidated information. The classification of indicators includes descriptions of system performance, capacity exceedance, and potential consequences of capacity exceedance. These classifications can also be categorized based on their relation to system events, including pre-event, intra-event, and post-event. Indicators can incorporate knowledge from both physical and social sciences into the decision-making process. Furthermore, they enable the assessment and adjustment necessary for achieving sustainable development goals.

Indicators can provide timely warnings to mitigate potential economic, social, and environmental setbacks. They also serve as effective tools for conveying ideas, thoughts, and values. Three indicators are especially relevant to climate change:

1. Reliability indicators
    - Hydraulic reliability index
    - Temporal reliability index
    - Volumetric reliability index
2. Resilience indicators
3. Sustainability indicators

6.9 CHAPTER PROBLEMS

1. What is a CBA, and what is its purpose?
2. Define the following key terms of CBA:
    - Direct and indirect costs
    - Intangible costs
    - Opportunity costs
    - Costs of potential risks
    - Present value and NPV of the project
3. Describe the process of conducting a CBA and its significance in decision making.
4. What methodology is used to quantify and evaluate financial expenditures and benefits?
5. What are direct benefits and indirect benefits?
6. How is the accuracy of a CBA measured?
7. What is the difference between life-cycle analysis and CBA?
8. Selecting appropriate and relevant criteria is a crucial aspect of MCA. The criteria should effectively capture the potential impact of all short-listed options. They should also enable a proper evaluation and comparison of these alternative options to address the identified need or service requirement. What are the five fundamental categories of MCA applied for the purpose of criteria selection?
9. Describe in detail the use of various weighting methodologies in MCA.

10. Describe the use of a pairwise scoring approach in MCA.
11. The City of "New Normal" is currently seeking proposals for professional engineering consultant services. These services will encompass various tasks, such as conducting hydrologic and hydraulic analysis (#1–2), completing pre-final and final design work (#3), preparing construction drawings and specifications (#4), estimating construction costs (#5), preparing bid sheets (#6), and providing construction management tasks (#7) during the construction phase. These services are specifically required for stormwater management improvement projects and related infrastructure enhancements outlined in the City of New Normal's 2023–24 Fiscal Year Budget. These projects are individually referred to as Component Projects #1, #2, #3, #4, #5, #6, and #7. Please provide a CBA for all projects, considering the implementation of an integrated system of storm sewers, detention and retention facilities, ditches, catch basins, inlets, overland floodways, flood routes, and other flood control infrastructure. The objective is to effectively manage stormwater runoff and mitigate flooding within the city.
12. Compare life-cycle costs for three alternatives (Alternative #1, Alternative #2, Alternative #3) of a stormwater retrofit project considering low impact development insulation R-values to determine the most cost-effective solution over a 40-year life. Include initial investment cost, operations cost, maintenance and repair cost, replacement cost, and residual value.

SELECTED SOURCES AND REFERENCES

Ahmad, E. 2016. *Infrastructure Finance in the Developing World: Public Finance Underpinnings for Infrastructure Financing in Developing Countries.* GGGI and G24. https://www.g24.org/wp-content/uploads/2016/05/MARGGK-WP05.pdf.

Anvarifar, F. 2011. "A Methodology for Risk-Based Optimization of Urban Drainage Systems." UNESCO-IHE Institute for Water Education, Delft, the Netherlands.

Archer, D., et al. 2014. "Moving Towards Inclusive Urban Adaptation: Approaches to Integrating Community-Based Adaptation to Climate Change at City and National Scale." *Climate and Development* 6, no. 4: 345–356. http://dx.doi.org/10.1080/17565529.2014.918868.

Argue, J. R., ed. 2013. *Water Sensitive Urban Design: Basic Procedures for 'Source Control' of Stormwater. A Handbook for Australian Practice.* Urban Water Resources Centre, University of South Australia.

Arnbjerg-Nielsen, K., P. Willems, J. Olsson, S. Beecham, A. Pathirana, I. Bülow Gregersen, and V. T. V. Nguyen. 2013. "Impacts of Climate Change on Rainfall Extremes and Urban Drainage Systems: A Review." *Water Science and Technology* 68, no. 1: 16–28.

Azumah, S. B., W. Adzawla, A. Osman, and P. Y. Anani. 2020. "Cost-Benefit Analysis of On-Farm Climate Change Adaptation Strategies in Ghana." *Ghana Journal of Geography* 12, no. 1: 29–46.

Bakhshipour, A. E., J. Hespen, A. Haghighi, U. Dittmer, and W. Nowak. 2021. "Integrating Structural Resilience in the Design of Urban Drainage Networks in Flat Areas Using a Simplified Multi-Objective Optimization Framework." *Water* 13, no. 3: 269.

Bakhshipour, A. E., U. Dittmer, A. Haghighi, and W. Nowak. 2021. "Toward Sustainable Urban Drainage Infrastructure Planning: A Combined Multiobjective Optimization and Multicriteria Decision-Making Platform." *Journal of Water Resources Planning and Management* 147, no. 8: 04021049.

Balsells, M., B. Barroca, V. Becue, and D. Serre. 2015. "Making Urban Flood Resilience More Operational: Current Practice." *Proceedings of the Institution of Civil Engineers-Water Management* 168, no. 2: 57–65.

Barreto, W. 2012. *Multi-Objective Optimization for Urban Drainage Rehabilitation.* CRC Press/Balkema. ISBN 978-0-415-62478-7.

Batalini de Macedo, M., M. Nobrega Gomes Jr., T. R. Pereira de Oliveira, M. H. Giacomoni, M. Imani, K. Zhang, and E. M. Mendiondo. 2022. "Low Impact Development Practices in the Context of United Nations Sustainable Development Goals: A New Concept, Lessons Learned and Challenges." *Critical Reviews in Environmental Science and Technology* 52, no. 14: 2538–2581.

Batica, J. and P. Gourbesville. 2012. "A Resilience Measures Towards Assessed Urban Flood Management CORFU Project." 9th International Conference on Urban Drainage Modelling, Belgrade.

Bhattacharya, A. 2016. "Delivering on Sustainable Infrastructure for Better Development and Better Climate." The Brookings Institution. https://www.brookings.edu/articles/delivering-on-sustainab le-infrastructure-for-better-development-and-better-climate/.

Blanco-Londoño, S. A., P. Torres-Lozada, and A. Galvis-Castaño. 2017. "Identification of Resilience Factors, Variables and Indicators for Sustainable Management of Urban Drainage Systems." Dyna 84, no. 203: 126–133.

Bony, S., et al. 2015. "Clouds, Circulation and Climate Sensitivity." Nature Geoscience 8, no. 4: 261–268. http://dx.doi.org/10.1038/ngeo2398.

Bush, E. and D. S. Lemmen, eds. 2019. Canada's Changing Climate Report. Government of Canada, Ottawa, ON, 444.

Casal-Campos, A., S. M. Sadr, G. Fu, and D. Butler. 2018. "Reliable, Resilient and Sustainable Urban Drainage Systems: An Analysis of Robustness Under Deep Uncertainty." Environmental Science & Technology 52, no. 16: 9008–9021.

Cervigni, R., et al., eds. 2015. Enhancing the Climate Resilience of Africa's Infrastructure: The Power and Water Sectors. The World Bank. http://dx.doi.org/10.1596/978-1-4648-0466-3.

Chu, E. 2016. "Urban Climate Adaptation and the Reshaping of State–Society Relations: The Politics of Community Knowledge and Mobilization in Indore, India." Urban Studies 1, no. 17. http://dx.doi .org/10.1177/0042098016686509.

City of Copenhagen. 2015. Climate Change Adaptation and Investment Statement. https://kk.sites.itera.dk/apps/kk_pub2/pdf/1499_bUxCj govgE.pdf.

Climate-ADAPT. 2016. The Economics of Managing Heavy Rains and Stormwater in Copenhagen—The Cloudburst Management Plan. https://climate-adapt.eea.europa.eu/metadata/case-studies/the-ec onomics-of-managing-heavyrains-and-stormwater-in-copenhag en-2013-the-cloudburst-management-plan/#source.

CSA Group. 2019a. *Flood Resilient Design of New Residential Communities, CSA W204:19.*

CSA Group. 2021. *Prioritization of Flood Risk in Existing Communities, CSA W210.*

DEA and SANBI. 2016. *Strategic Framework and Overarching Implementation Plan for Ecosystem-Based Adaptation (EbA).*

Delelegn, S., A. Pathirana, B. Gersonius, A. Adeogun, and K. Vairavamoorthy. 2011. "Multi-Objective Optimization of Cost-Benefit of Urban Flood Management Using a 1 D 2 D Coupled Model." *Water Science and Technology* 63: 1054.

Dong, X., H. Guo, and S. Zeng. 2017. "Enhancing Future Resilience in Urban Drainage System: Green Versus Grey Infrastructure." *Water Research* 124: 280–289.

Dudley, S., et al. 2017. "Consumer's Guide to Regulatory Impact Analysis: Ten Tips for Being an Informed Policymaker." *Journal of Benefit-Cost Analysis* 8, no. 2: 187–204. https://doi.org/10.1017/bca.2017.11.

Farmani, R., G. A. Walters, and D. A. Savic. 2005. "Trade-Off Between Total Cost and Reliability for Anytown Water Distribution Network." *Journal of Water Resources Planning and Management* 131, no. 3: 161–171.

Field, C., V. Barros, T. F. Stocker, D. Qin, D. J. Dokken, K. L. Ebi, et al. 2012. *Managing the Risks of Extreme Events and Disasters to Advance Climate Change Adaptation.* IPCC Special Report. Cambridge University Press.

Fletcher, T. D., H. Andrieu, and P. Hamel. 2013. "Understanding, Management and Modelling of Urban Hydrology and Its Consequences for Receiving Waters: A State of the Art." *Advances in Water Resources* 51: 261–279.

Florin, M.-V. and I. Linkov, eds. 2016. *IRGC Resource Guide on Resilience.* Lausanne: EPFL International Risk Governance Center (IRGC).

Forzieri, G., et al. 2018. "Escalating Impacts of Climate Extremes on Critical Infrastructures in Europe." *Global Environmental Change* 48: 97–107. http://dx.doi.org/10.1016/J.GLOENVCHA.2017.11.007.

Fratini, C. F., G. D. Geldof, J. Kluck, and P. S. Mikkelsen. 2012. "Three Points Approach (3PA) for Urban Flood Risk Management: A Tool to Support Climate Change Adaptation Through Transdisciplinarity and Multifunctionality." *Urban Water Journal* 9, no. 5: 317–331.

GHD. 2020. *Climate Data for Hydrologic and Hydraulic Analysis, Technical Memorandum #3: IDF Curves.*

Government of Canada. 2015. *Short-Duration Rainfall Intensity-Duration-Frequency.*

Guptha, G. C., S. Swain, N. Al-Ansari, A. K. Taloor, and D. Dayal. 2021. "Evaluation of an Urban Drainage System and Its Resilience Using Remote Sensing and GIS." *Remote Sensing Applications: Society and Environment* 23: 100601.

Hager, W. H. 2010. *Wastewater Hydraulics: Theory and Practice.* Berlin: Springer.

Hammond, M., A. Chen, S. Djordjević, D. Butler, and O. Mark. 2013. "Urban Flood Impact Assessment: A State-of-the-Art Review." *Urban Water Journal* 1–16.

Hariri-Ardebili, M. A. 2018. "Risk, Reliability, Resilience (R3) and Beyond in Dam Engineering: A State-of-the-Art Review." *International Journal of Disaster Risk Reduction* 31: 806–831.

Hisschemöller, M. and E. Cuppen. 2015. "Participatory Assessment: Tools for Empowering, Learning and Legimating." In *The Tools of Policy Formulation: Actors, Capacities, Venues and Effects*, edited by A. Jordan and J. R. Turnpenny. Edward Elgar, Cheltenham.

Index, C. R. 2014. *City Resilience Framework.* The Rockefeller Foundation and ARUP.

Islam, N. and R. Mechler. 2007. *Orchid: Piloting Climate Risk Screening in DFID Bangladesh: An Economic and Cost Benefit Analysis of Adaptation Options.* Sussex, UK, Institute of Development Studies.

Jha, A., T. Todd, and Z. Stanton-Geddes. 2013. *Building Urban Resilience: Principles, Tools and Practice.* International Bank for Reconstruction and Development / The World Bank. ISBN 978-0-8213-9826-5.

Jordan, A. and J. R. Turnpenny, eds. *The Tools of Policy Formulation: Actors, Capacities, Venues and Effects.* Edward Elgar, Cheltenham.

Kerr Wood Leidal Associates Ltd. 1999. *Stormwater Management Strategy for Mission/Wagg Creek System.* Report to City of North Vancouver, BC.

Lacambra Ayuso, S., L. Wolfram, and C. Baubion, eds. 2017. *Policy Evaluation Framework on the Governance of Critical Infrastructure Resilience in Latin America.* Inter-American Development Bank. http://dx.doi.org/10.18235/0000819.

Lamond, J. E., C. B. Rose, and C. A. Booth. 2015. "Evidence for Improved Urban Flood Resilience by Sustainable Drainage Retrofit." *Proceedings of the Institution of Civil Engineers-Urban Design and Planning* 168, no. 2: 101–111.

Liu, L., W. Y. Shao, and D. Z. Zhu. 2020. "Experimental Study on Stormwater Geyser in a Vertical Shaft Above a Junction Chamber." *ASCE Journal of Hydraulic Engineering* 146.

Liu, W. and Z. Song. 2020. "Review of Studies on the Resilience of Urban Critical Infrastructure Networks." *Reliability Engineering & System Safety* 193: 106617.

Ma, Y. Y., D. Z. Zhu, N. Rajaratnam, and B. van Duin. 2017. "Energy Dissipation in Circular Drop Manholes." *ASCE Journal of Irrigation and Drainage Engineering* 143: 04017047.

Mannina, G. and G. Viviani. 2010. "An Urban Drainage Stormwater Quality Model: Model Development and Uncertainty Quantification." *Journal of Hydrology* 381, no. 3–4: 248–265.

Martin, C., Y. Ruperd, and M. Legret. 2007. "Urban Stormwater Drainage Management: The Development of a Multicriteria Decision Aid Approach for Best Management Practices." *European Journal of Operational Research* 181, no. 1: 338–349.

Martínez-Cano, C., B. Toloh, A. Sanchez-Torres, Z. Vojinović, and D. Brdjanovic. 2014. "Flood Resilience Assessment in Urban Drainage Systems Through Multi-Objective Optimisation."

Matungulu, H. 2010. "Comparison of Different Urban Flood Modelling Approaches Within the Context of Optimization of Rehabilitation Measures." UNESCO-IHE Institute for Water Education, Delft, the Netherlands.

Meyer, M., M. Flood, J. Keller, J. Lennon, G. McVoy, C. Dorney, et al. 2014. *Volume 2: Climate Change, Extreme Weather Events and the Highway System: Practitioner's Guide and Research Report*. NCHRP Report 750, Transportation Research Board, Washington, D.C.

Mishan, E. J. and E. Quah. 2020. *Cost-Benefit Analysis*. Routledge.

National Research Council. 2012. *Disaster Resilience: A National Imperative*. The National Academies Press, Washington, D.C.

Noleppa, S. 2013. *Economic Approaches for Assessing Climate Change Adaptation Options Under Uncertainty: Excel Tools for Cost-Benefit and Multi-Criteria Analysis*, GIZ.

Owusu, K. and P. B. Obour. 2020. "Urban Flooding, Adaptation Strategies, and Resilience: Case Study of Accra, Ghana." *African Handbook of Climate Change Adaptation* 1–17.

Panos, C. L., J. M. Wolfand, and T. S. Hogue. 2021. "Assessing Resilience of a Dual Drainage Urban System to Redevelopment and Climate Change." *Journal of Hydrology* 596: 126101.

Paredes Méndez, D. F., A. Sanchez-Torres, Z. Vojinović, and S. D. Seyoum. (2014). Multi-Objective-Rehabilitation of Urban Drainage Systems Within the Flood Risk Framework. UNESCO-IHE Institute for Water Education, Delft, the Netherlands.

Pérez-Soba, M. and R. Maas. 2015. "Scenarios: Tools for Coping with Complexity and Future Uncertainty?" In *The Tools of Policy Formulation: Actors, Capacities, Venues and Effects*, edited by A. Jordan and J.R. Turnpenny. Edward Elgar, Cheltenham.

Public Infrastructure Engineering Vulnerability Committee (PIEVC). 2020. *PIEVC Program*. https://pievc.ca.

Qian, Y., D. Z. Zhu, L. Liu, W. Y. Shao, S. Edwini-Bonsu, and F. Zhou. 2020. "Numerical and Experimental Study on Mitigation of Storm Geysers." *ASCE Journal of Hydraulic Engineering* 146. doi: 10.1061/ (ASCE)HY.1943-7900.0001684.

Rezende, O. M., F. M. Miranda, A. N. Haddad, and M. G. Miguez. 2019. "A Framework to Evaluate Urban Flood Resilience of Design Alternatives for Flood Defence Considering Future Adverse Scenarios." *Water* 11, no. 7: 1485.

Saarikoski, H., J. Mustajoki, D. N. Barton, D. Geneletti, J. Langemeyer, E. Gomez-Baggethun, et al. 2016. "Multi-Criteria Decision Analysis and Cost-Benefit Analysis: Comparing Alternative Frameworks for Integrated Valuation of Ecosystem Services." *Ecosystem Services* 22: 238–249.

Sartori, D., G. Catalano, M. Genco, C. Pancotti, E. Sirtori, S. Vignetti, and C. Bo. 2014. *Guide to Cost-Benefit Analysis of Investment Projects. Economic Appraisal Tool for Cohesion Policy 2014–2020.*

Savic, D. 2008. "Global and Evolutionary Optimization for Water Management Problems." In *Practical Hydroinformatics Computational Intelligence and Technological Developments in Water Applications*, edited by R. Abrahart, L. See, and D. Solomatine. ISBN: 978-3-540-79880-4.

Sayers, P., Y. G. Li, E. Galloway, F. Penning-Rowsell, K. Shen, Y. Wen, et al. 2013. *Flood Risk Management: A Strategic Approach.* Paris: UNESCO.

Schütze, M., D. Butler, and M. Beck. 2002. *Modelling, Simulation and Control of Urban Wastewater Systems.* Springer: London.

Seyoum, S., Z. Vojinovic, R. Price, and S. Weesakul. 2012. "A Coupled 1D and Non-Inertia 2D Flood Inundation Model for Simulation of Urban Flooding." *ASCE Journal of Hydraulic Engineering* 138, no. 1: 23–34.

Sharifi, A. 2020. "Urban Resilience Assessment: Mapping Knowledge Structure and Trends." *Sustainability* 12, no. 15: 5918.

———. 2021. "Co-Benefits and Synergies Between Urban Climate Change Mitigation and Adaptation Measures: A Literature Review." *Science of the Total Environment* 750: 141642.

Siekmann, T. and M. Siekmann. 2015. "Resilient Urban Drainage–Options of an Optimized Area-Management." *Urban Water Journal* 12, no. 1: 44–51.

Tang, Y. B., D. Z. Zhu, N. Rajaratnam, and B. van Duin. 2020. "Sediment Depositions in a Submerged Storm Sewer Pipe." *ASCE Journal of Environmental Engineering* 146. doi: 10.1061/(ASCE) EE.1943-7870.0001799.

Tian, W., Z. Liao, Z. Zhang, H. Wu, and K. Xin. "Flooding and Over-flow Mitigation Using Deep Reinforcement Learning Based on Koopman Operator of Urban Drainage Systems." *Water Resources Research* e2021WR030939.

UK NEA. 2011. *National Ecosystem Assessment: Technical Report*. UN-EP-WCMC, Cambridge.

Urban Drainage Flood Control District. 2016. *Urban Storm Drainage Criteria Manual: Volume 1 Management, Hydrology, and Hydraulics*. Denver, CO.

van Duin, B., D. Z. Zhu, W. Zhang, R. J. Muir, C. Johnston, C. Kipkie, and G. Rivard. 2021. "Toward More Resilient Urban Stormwater Management Systems—Bridging the Gap from Theory to Implementation." *Frontiers in Water* 3: 671059.

Vélez, C. 2012. *Optimization of Urban Wastewater Systems Using Model Based Design and Control*. CRC Press/Balkema. ISBN 978-1-138-00002-5.

Vojinovic, Z., A. Sánchez, and W. Barreto. 2008. "Optimising Sewer System Rehabilitation Strategies Between Flooding, Overflow Emissions and Investment Costs." 11th International Conference on Urban Drainage, Edinburgh, Scotland, UK.

Walesh, S. G. 1999. *Street Storage System for Control of Combined Sewer Surcharge—Retrofitting Stormwater Storage Into Combined Sewer Systems*. National Risk Management Research Laboratory, U.S. EPA, Cincinnati, OH.

Wang, M., Y. Fang, and C. Sweetapple. 2021. "Assessing Flood Resilience of Urban Drainage System Based on a 'Do-Nothing' Benchmark." *Journal of Environmental Management* 288: 112472.

Wang, M., Y. Zhang, A. E. Bakhshipour, M. Liu, Q. Rao, and Z. Lu. 2022. "Designing Coupled LID–GREI Urban Drainage Systems: Resilience Assessment and Decision-Making Framework." *Science of the Total Environment* 834: 155267.

Wang, Y., A. O. Rousis, and G. Strbac. 2021. "A Three-Level Planning Model for Optimal Sizing of Networked Microgrids Considering a Trade-Off Between Resilience and Cost." *IEEE Transactions on Power Systems* 36, no. 6: 5657–5669.

Watt, E. W., ed. 1989. *Hydrology of Floods in Canada: A Guide to Planning and Design*. National Research Council of Canada, Ottawa, ON.

Wilby, R. L. and R. Keenan. 2012. "Adapting to Flood Risk Under Climate Change." *Progress in Physical Geography* 36, no. 3: 348–378.

Wisner, P. E. and A. M. Kassem. 1980. "Street Overland Flow and Inlet Control, Annual Conference." Winnipeg, MB: Canadian Society of Civil Engineering.

Wisner, P. E., A. M. Kassem, and P. W. Cheung. 1981. "Parks Against Storms Proceedings." In *Second International Conference on Urban Storm Drainage*. Urbana, IL: 322–330.

Yin, X. H., G. Y. Hao, and F. Sterck. 2022. "A Trade-Off Between Growth and Hydraulic Resilience Against Freezing Leads to Divergent Adaptations Among Temperate Tree Species." *Functional Ecology* 36, no. 3: 739–750.

Zarghami, S. A. and I. Gunawan. 2021. "Forecasting the Impact of Population Growth on Robustness of Water Distribution Networks: A System Dynamics Approach." *IEEE Transactions on Engineering Management*.

Zhou, Q. 2014. "A Review of Sustainable Urban Drainage Systems Considering the Climate Change and Urbanization Impacts." *Water* 6, no. 4: 976–992.

Zischg, J. 2018. "Understanding Patterns, Dependencies and Resilience in Complex Urban Water Infrastructure Networks."

INDEX

Page numbers followed by "*f*" and "*t*" refer to figures and tables, respectively.